Ana Cláudia Caminha de Melo
Elvira Maria G. Seixas Maciel
Elisabeth Regina Alves C. Silva

Occupation, land use and food security

Ana Cláudia Caminha de Melo
Elvira Maria G. Seixas Maciel
Elisabeth Regina Alves C. Silva

Occupation, land use and food security

Agribusiness vs. family farming

ScienciaScripts

Imprint

Any brand names and product names mentioned in this book are subject to trademark, brand or patent protection and are trademarks or registered trademarks of their respective holders. The use of brand names, product names, common names, trade names, product descriptions etc. even without a particular marking in this work is in no way to be construed to mean that such names may be regarded as unrestricted in respect of trademark and brand protection legislation and could thus be used by anyone.

Cover image: www.ingimage.com

This book is a translation from the original published under ISBN 978-613-9-66406-1.

Publisher:
Sciencia Scripts
is a trademark of
Dodo Books Indian Ocean Ltd. and OmniScriptum S.R.L publishing group

120 High Road, East Finchley, London, N2 9ED, United Kingdom
Str. Armeneasca 28/1, office 1, Chisinau MD-2012, Republic of Moldova, Europe
Printed at: see last page
ISBN: 978-620-8-02765-0

I dedicate this work to the people most present in my life:
My mum, the most generous and giving of all mums;
My sisters, for being by my side at every moment of my life;
My husband, Ricardo, for his companionship and encouragement;
My daughter, Giovana, my greatest GIFT!
To God, who gives me everything.

ACKNOWLEDGEMENTS

Firstly to God, who allowed all this to happen and who at all times is the greatest teacher anyone can have;

To my family for teaching me never to give up, for always believing in me;

To my mother, Raimunda, and my father, Danúnzio (in *memoriam)*, who have always believed in education as a form of liberation and independence;

To my friend and beloved husband, Ricardo Hortegal, for always being by my side and encouraging me;

To my beloved daughter, Giovana, who was able to understand my absences during my master's programme;

To my sister, Valéria, for her willingness to drive me to the interviews, giving up her activities;

To my supervisor, Elvira Maciel, who believed in my potential in a way that I didn't think I could match;

To the Federal Institute of Education, Science and Technology of Maranhão (IFMA), which made this master's degree possible;

To Professor Sérgio Koifinan (in *memoriam)*, who made it possible for IFMA staff to take part in the Public Health and Environment programme through an inter-institutional master's degree. I am extremely grateful;

To the professors and coordinators of the master's programme, Rosalina Koifman and Gina Torres, for their technical excellence and sensitivity in conducting the work in search of quality training;

To Jeovani Rodrigues, for his partnership and help in preparing and carrying out the work;

To Elisvanda Ramos and Tamiles Borges for giving up their time to help me carry out the field research;

To IFMA-Campus Maracanã, especially Cledes Carneiro, for the time given, reception, collaboration and essential help in carrying out this work;

To my master's degree friends, who were part of this whole journey, for the great times and great help;

The local coordination, especially Graça Sampaio, for helping to organise this course so that it could be a success;

To all the professors on the programme who came to São Luís to share their knowledge with us.

Every education system is a political way of
maintaining or modifying the appropriation of
discourses, with the knowledge and powers they
bring with them.

Michel Foucault

SUMMARY

Food insecurity, especially when expressed through hunger, is prevalent among people living in poverty. This complex phenomenon, which currently affects almost a million people worldwide, can be explained by various factors, with difficulties in accessing food being one of the main causes. The vast majority of people who are food insecure due to lack of access to food belong to the rural population. Since agricultural schools were designed to serve the working class, they now have to prioritise socio-economic and environmental development, given that for a long time they were focused on serving large agribusiness companies. Knowing how teachers at an agricultural school think and teach about the relationship between food security, land use and occupation, and how this affects the health of the population is important, as it has a direct effect on the students' education. Bearing in mind that students will be future professionals, the aim of this study was to uncover the knowledge and perceptions of agricultural students about the ways in which farmers and family farmers produce and their consequences for the population. To this end, an exploratory study was carried out using individual interviews and focus groups with students regularly enrolled in the subsequent technical course in agriculture at the Federal Institute of Education, Science and Technology of Maranhão-Campus Maracanã. Data collection was preceded by a literature review, followed by field research with semi-structured interviews and focus groups with students and teachers. The descriptive analysis of the data was carried out using the Collective Subject Discourse technique, which is a technique for organising the material resulting from fieldwork, usually from interviews. We realised that the professional training of agricultural technicians left something to be desired in terms of content that addresses the issue of food security with a focus on the social impacts caused by the predominant agricultural production model in Brazil. In this way, it hinders future professionals from taking a critical view of this production model, which generates social inequalities and could jeopardise their role as rural development agents for family farmers.

Key words: Agricultural Technician. Food security. Occupation. Land Use.

SUMMARY

CHAPTER 1

INTRODUCTION

The world today celebrates its great potential in food production. The world currently has the capacity to provide food for twelve billion people, with nearly seven billion inhabitants, but of these, almost one billion human beings suffer from permanent hunger (ZIEGLER, 2013).

One of the main factors responsible for the hunger of a significant number of people is the current economic model, which focuses on high production at all costs, disregarding environmental and social aspects. In the agricultural sector in particular, the transformation of traditional agriculture into modern agriculture has led to the occupation of small properties by monoculture, a decrease in the practice of family farming and a qualitative and quantitative transformation of labour opportunities in rural areas. This has led, in certain regions, to an increase in poverty and, consequently, food insecurity and hunger.

In Brazil, another trigger for poverty is the inequality resulting from poor income distribution. The current development model, which favours macroeconomic indicators, relegates social indicators to the background. Health problems linked to malnutrition, for example, have irreversible consequences, especially in childhood. Social Determinants of Health (SDH) are the factors that lead people to have different health conditions as a result of social, economic and environmental factors.

It is well known that population groups have inequalities when it comes to health. Among these, some inequalities are so-called inevitable differences, because they are linked to biological aspects such as gender, age, etc. However, when these inequalities are the result of modes of organisation that are chosen, we can say that they are unnecessary and avoidable. Inequity" refers to avoidable inequalities. Therefore, for there to be favourable health conditions in the various population groups, equity must be sought, defined as the willingness to recognise the right to adequate living conditions for everyone in society, to meet different demands differently, in order to promote social justice. The SDH must be recognised by public policies as an object of action since, according to the 1988 Federal Constitution, "health is a citizen's right and a duty of the state".

United Nations bodies and various non-governmental organisations set out to meet food needs (and their consequences) in very poor areas. Brazil has been carrying out social programmes aimed at eradicating poverty in the country. In the contemporary scenario, there is no denying that there are discussions, different initiatives and the development of actions aimed at the problem both in Brazil and around the world. However, these discussions are not present in the complexity of life, in the day-to-day life of society. Hunger sometimes seems to represent something natural; inevitably, on an increasingly populated planet, some would have more and others would have less (or nothing at all).

In the consumer society, the concern with profitability in production and commercialisation, as well as the accumulation of goods, has been greater than with attempts to distribute income less unequally and to offer essential goods and services, if not to everyone, to a greater number of people. It's about extending citizenship. Through formal or informal education and the assimilation of information disseminated by the mass media, what can be observed is a certain naturalisation in the face of everyday facts and events that the individual witnesses with a certain frequency, but whose perception does not lead him or her to wonder or question them because they don't bother him or her.

The proposal for this work arises from the assumption that there is a group, with which I am involved in my teaching activity, that should not be unaware of problems relating to the environmental and social effects that agricultural activity geared towards agribusiness has caused in the interior of the country: this group is made up of professionals linked to the agricultural sector, be they agricultural technicians or those who train these professionals.

At a time in history when Brazil and the world are discussing the search for values such as sustainability, considering that the current form of production has proven to be degrading in terms of the exploitation and use of natural resources, the preservation of ecosystems and mankind, the importance of the role of education must be emphasised in this scenario. When based on the critical formation of citizens, it works as a resource capable of transforming mentalities, bringing the habit of problematising certain situations into citizens' daily lives. Through discussions, knowledge of a propositional nature is built up - if not to solve the problem as a whole, at least to control it in more appropriate ways.

Education, particularly technical vocational education, seeks to train professionals to work in areas that are in high demand on the labour market. The agricultural market has been growing, and in a context of agribusiness expansion, the ways in which land is used for agriculture in the country has undergone transformations whose socio-environmental impacts are not always positive. It is for this changing field that technicians should be prepared, including information and reflection on the social, political and economic aspects involved.

The assumption of this dissertation is that education is needed that focuses on the critical training of professionals, i.e. during the educational process, elements should be provided that favour the development of a vision that is less centred on technique and leads them to see and interpret the world beyond the logic of the labour market, seeking to make individuals capable of thinking and acting responsibly in order to promote practices aimed at improving the quality of life (GOMES; MARTINS, 2004).

Since the Federal Institutes of Education, Science and Technology are the most renowned institutions in the country in terms of technical vocational training, meeting much of the demand from the industrial and agricultural markets, it is important to know how the training of technicians for the agricultural sector is going, since they will be involved in a wide range of activities in the sector, often guiding the decision-making of rural workers, especially smallholders and family farmers.

One of the main aims of technical education in today's world should be to make professionals aware of and reflect on the reality that surrounds them, especially the reality constructed from the activities carried out by the production sector in which they are or will soon be working. This will provide them with elements that will enable them to make assessments and decisions more consciously and responsibly.

Focussing on agricultural vocational training in the state of Maranhão, which has always stood out as one of the states in the country with the lowest Human and Social Development Index (HDI), this paper aims to reflect on the vocational training offered by the Federal Institute of Maranhão (IFMA), seeking, by studying the agricultural training segment at the IFMA - Maracanã Campus, the need and possibility of integrating educational disciplines and foundations into the school curriculum. These changes should enable and encourage students and teachers to reflect on the relationship between the way land is occupied and used (expansion of agribusiness, intensive use of pesticides, encouraging family and organic farming) and the population's food security.

CHAPTER 2

THEORETICAL FRAMEWORK

2.1 Agribusiness, family farming and food security: economic development and socio-environmental issues

It is not difficult to understand that agricultural production, as part of a certain economic development model, determines the ways in which land is occupied and used, acting as a determining factor in the social and environmental conditions that are established in the territory.

Agribusiness is characterised as a form of production geared towards exports. Its characteristic is the concentration of land use - large tracts of land destined for the same type of crop. In contrast, family farming is carried out on small properties, its production is geared towards the local market and it guarantees access to healthy food for many Brazilians. It also emerges that rural areas absorb a considerable amount of labour, but depending on the form of production, family or employer, there is a discrepancy in the number of workers employed.

Therefore, since agribusiness has the largest land occupation and the smallest number of rural labourers, the social effects caused by this form of production can be estimated (FREITA; GARCIA, 2012). These social effects stem from the fact that agribusiness occupies a lot of land, but generates a small number of jobs, and these jobs are largely precarious. As for the environmental effects caused by this form of production, we note the use of a large quantity of chemical inputs (agrotoxins or agrochemicals) and this has a major effect on workers' health.

This is the reality of agribusiness in many places, which reinforces the idea that territorial organisation is a historical and social product of the relationship between society and nature, mediated by technology. The technique used, which depends on the type of economic activity, introduces new relationships with space, with social, cultural, political and environmental implications (SANTOS, 2002).

Having drawn some basic distinctions between these forms of production, we now turn to some concepts of family farming and agribusiness. Family farming can be conceptualised from the perspective of Brazilian legislation. According to Law 11.326/2006, a property with 4 fiscal modules and the hiring of up to two permanent employees are considered family farms. It also stipulates that labour must predominantly come from the family itself and income must be derived from the property's activities; management must also be carried out by a family member.

The term *"agribusiness"* was coined in the USA in 1955 to emphasise the growing interrelationship between agriculture, industry and services. In Brazil, the concept emerged in the 1980s with the expression "Agroindustrial Complex", which later evolved into agribusiness. It is well known that the construction of concepts is subject to ideological disputes, in this case over what represents Brazilian agriculture. Based on this understanding (of the concept of agribusiness), there is a tendency to differentiate family farming from agribusiness based on scale, so the differences in these forms of production end up being summarised in the size of production, which is why the terms "agribusiness" are commonly used, agribusiness and family agribusiness (agronegocinho), but for the issue addressed here, these differences in forms of production are not limited to scale, because "it is not what is done, but how, with what means of labour it is done", their socio-environmental implications are what will actually determine these differences.

In recent years, agribusiness in Brazil has experienced growth that has enabled it to play a significant role in the country's economic development. One of its contributions has been to the Gross Domestic Product (GDP), a fact that has favoured high levels of investment by the Brazilian government. With the development of the sector, Brazil has stood out on the international market as a major food producer and is expected to increase production.

In addition to investments in the sector favouring agricultural development, the Food and Agriculture Organization of the United Nations (FAO) is projecting an increase in income for several countries by 2050, which will increase the demand for food, favouring greater economic growth for Brazil.

However, the large availability of food in Brazil and other countries, in addition to the FAO projections mentioned above, does not guarantee access to food for everyone. One determining factor that gives countries, communities and individuals access to food is their purchasing power, i.e. their income. However, social inequality occurs strongly between countries and within countries. A clear example of this is Brazil, where the average monthly per capita income available to poor people was R$ 43.09 in 2001, which shows a great difference between social classes in the country and, consequently, differences in access to food (BELIK, 2003). Social indicators have improved in recent years, but the gap between the extremes is still very wide.

According to Schmidhuber (2011), the availability of food is of no use if there are no monetary or non-monetary resources to ensure equal food and nutritional security for the population. Food and Nutrition Security (FNS) is understood as: the realisation of everyone's right to regular and permanent access to quality food, in sufficient quantity, without compromising access to other essential needs, based on health-promoting eating practices that respect cultural diversity and are socially, economically and environmentally sustainable. In order to guarantee this right, there is a need for public policies (CONSEA, 2004).

According to Burlandy (2007), one of the causes of poverty in Brazil is the unequal distribution of income, which leads to food and nutritional insecurity. Considering this factor, we can understand the contribution of agribusiness to its worsening, since it is characterised by the concentration of land and the intensification of agricultural mechanisation, both of which make farmers unemployed and drive them out of the countryside. As a result, a large number of people live in situations of risk and vulnerability.

The determination of a situation of food insecurity is directly related to this situation of conflict and exclusion in the countryside. Of the huge contingent of 30 million people who suffer from hunger in this country, half are in rural areas, although only 21 per cent of its total population lives in the countryside. Recent studies have revealed that in agrarian reform settlements in Brazil, which have already achieved a reasonable degree of consolidation, the nutritional status found is even higher than that enjoyed by conventional family farming. On the other hand, in landless encampments in occupied areas, which have not yet been recognised and are therefore not in a position to produce, the malnutrition situation is extremely serious. (MALUF, 2013)

Over the last ten years, the Brazilian government, through the "Zero Hunger" conditional cash transfer programme, has sought to end hunger and poverty in the country. In Brazil, FNS is a human right guaranteed in the Food and Nutrition Security Law passed in 2006. However, it is true that just transferring income to those who receive it in the form of a grant, for many reasons, does not guarantee the stability of FNS. Hence the need for integrated projects to help make the programme's objectives a reality (BURLANDY, 2007).

One of the ways to increase food access and social equity would be to encourage family farming. Family farming creates more jobs, reduces the distance between producer and consumer, makes food cheaper and therefore more accessible, increases the diversity of food produced and favours local eating habits.

For a long time, family farming was seen as a backward form of production and incapable of promoting economic development for the country, so the modernisation of the latifundia was seen as the great way out of economic growth and the guarantee of large-scale production. However, unlike some capitalist countries, Brazil promoted economic growth based on the exclusion of family farming. Today, the countries with the best human development indices, such as the United States and Japan, have promoted their development with the strong presence of family farming, resulting in dynamic economies and more democratic and equitable societies (GUANZIROLI, 2009).

Family farming keeps people in the countryside and has more sustainable production bases in terms of environmental and health issues. What's more, it is now known that family farming, with technical guidance and support for agricultural credits, i.e. government support, has the capacity to increase production, revealing its economically viable and socially just nature, guaranteed by better income distribution.

In this way, the country's current food production potential could be combined with the ease of access to food provided by family farming, promoting food and nutritional security for many who have been put at risk by labour-saving forms of agricultural production.

We can see, then, that the consensus of the supremacy of agribusiness over family farming has failed, because what we see today in Brazil is a situation of social inequality and poverty that has led the state to intervene with conditional cash transfers because it itself has not taken into account for a long time the power of agribusiness to generate inequality. It can be seen, then, that the meaning of development does not only favour economic factors to the detriment of the human condition, but also values that promote social equity (SANTOS, 2002). In this sense, in order to identify a country's level of development, the Human Development Index (HDI) is more efficient than the Gross Domestic Product (GDP).

Since the 1990s, much has been discussed about rural development, and statistical data from the time, such as the PNAD, revealed that the active labour force in rural areas was decreasing with the entry of agribusiness. This evidence only reinforced the idea that agribusiness was a form of production with economic potential for the country, but one that generated a lot of social inequalities and, consequently, poverty. Meanwhile, various studies on family farming have strengthened its potential for local economic development, as well as providing social equity (SCHNEIDER, 2010).

By organising the ideas of economic development with social equity, it can be understood that family farming is capable of promoting local development, which is understood as:

> Local development, on the other hand, is part of this systemic vision of development and can be defined as: an endogenous process registered in small territorial units and human groupings capable of promoting economic dynamism and improving the population's quality of life. It represents a singular transformation in the economic bases and social organisation at local level, resulting from the mobilisation of society's energies by exploiting its specific capacities and potential (BUARQUE, 1998).

From the text above, it is clear that family farming is important for promoting development in a way that includes social equity. Another aspect that favours the strengthening of the practice of family farming has been the numerous questions about the Green Revolution:

> Green Revolution - a programme spread by the Americans that used new technologies based on plant genetics, the creation and multiplication of seeds resistant to diseases and pests, as well as modern and efficient agricultural techniques to increase productivity and grain production (BARBOSA, 2010).

Among these questions was the end of world hunger, which was not promised by this programme. These questions have led us to think about new forms of production that guarantee greater income distribution and food security.

According to Ziegler (2013): "The planet is saturated with wealth. So there is no fatality. And if a billion people suffer from hunger, it's not because of poor food production, but because of the hoarding of the fruits of the earth by the most powerful."

In addition to the problem of hunger, many studies have now revealed the high environmental impacts of agribusiness. Monocultures, causing soil erosion, indiscriminate and excessive use of water, use of transgenic seeds, contamination of soil and water resources by pesticides and others.

As for pesticides, they pose a great risk to human beings and can be found in food, either directly during the production process or indirectly, as in the case of beef animals fed contaminated vegetable feed (FARIA, 2003).

Considering the concept of FNS according to the Organic Law of 2006, which provides for the adoption of health-promoting dietary practices for quality food, the use of chemicals in food production that pose a risk to human beings is outside the scope of the issue of safe, quality food. According to the FAO, pesticides are defined as:

> Any substance, or mixture of substances, used to prevent, destroy or control any pest - including vectors of human and animal diseases, unwanted species of plants or animals, causing damage during (or interfering with) the production, processing, storage, transport or distribution of food, agricultural products, timber and derivatives, or which is to be administered for the control of insects, arachnids and other pests affecting the bodies of farm animals (FAO, 2003).

In this sense, Brazil's situation is worrying, as it is a major food producer and also the country that has been using the most pesticides in agriculture for some years. The danger of this is revealed by the higher rates of pesticides in food, as well as the contamination of soil, rivers and groundwater (PERES; MOREIRA, 2007).

The wealth of food production on the planet today has led to a compromised quality of life:

> Humanity is living through a moment it has never experienced before: it is seen as a whole, rich and powerful. But on the other hand, no one can believe that this accumulation of power can continue unabated... without turning in on itself and without threatening the physical and moral survival of humanity (CONVIVIALIST MANIFESTO, 2013).

When it comes to the relationship between food security and health, we need to bear in mind what is meant by health. The World Health Organisation (WHO) defines health as a state of complete physical, mental and social well-being, and not merely the absence of disease or infirmity. This concept clearly expresses a very broad conception of health, going beyond the disease-centred approach to also include factors that can interfere with the state of health (BELIK, 2003). Currently, these factors are called Social Determinants of Health (SDH), defined as social, economic, cultural, ethnic/racial, psychological and behavioural factors that influence the occurrence of health problems and their risk factors for the population (BRASIL, 2006).

Based on the concepts discussed above, an example of an SDH is income. If income distribution is not equal within a country, there tends to be different health situations within population groups living in the same urban centre, for example. This is due to the difference in access to adequate food, leading people with a low *per capita* family income to be exposed to nutritional deficiencies. It can therefore be seen that these are not genetic-biological factors - they are therefore possible to avoid.

When it comes to health in relation to food security, the wake-up call for these discussions in Brazil was given by the works of Josué de Castro, a Brazilian doctor, who revealed the problem of FNS as the result of the economic and social development model, with poor nutrition stemming from the unequal distribution of income (BRASIL, 2006).

In Brazil, the issue of FNS is often discussed from the point of view of poverty and misery, although this concept is currently not limited to combating hunger, but includes the search for healthy food (PINHEIRO, 2008). This is because poverty is one of the main factors leading to food insecurity in the country, revealing inequality in income distribution as one of the obstacles that need to be tackled by public policies.

Seeking to minimise the problem of hunger in the country, Brazil signed a commitment at the World Food Summit in 1996, along with other countries, to halve the number of hungry people by 2015. One action that has helped the country fulfil this commitment was the creation of the National Food and Nutrition Security Policy (PNSAN), whose instrument is the National Food and Nutrition Security Plan (PLANSAN). In order to guarantee, through intersectoral actions, the Human Right to Adequate Food (DHAA), the National Food and Nutrition Security System (SISAN) was created by the Organic Food Security Law (LOSAN) - Law^0 11.346 of 2006. Its principles for implementing the National Food and Nutrition Security Policy are social participation and intersectorality (BRASIL, 2003).

According to Brasil (2006) it is understood that:

> Food and Nutrition Security (FNS) encompasses the realisation of everyone's right to regular and permanent access to quality food, in sufficient quantity, without compromising access to other essential needs, based on health-promoting dietary practices that respect cultural diversity and are environmentally, culturally, economically and socially sustainable.

The principles of food sovereignty and sustainability have been incorporated into this idea of FNS. The former is related to the culture and eating habits of each country and sustainability incorporates concepts linked to environmental preservation, condemning the use of pesticides and extensive monoculture production (BELIK, 2003). The relationship between FNS and patterns of food production and consumption can therefore be seen to be intimate.

> The model of food production and consumption is fundamental to guaranteeing food security because, in addition to hunger, there is food and nutritional insecurity whenever food is produced without respect for the environment, with the use of pesticides that affect the health of workers and consumers, without respect for the precautionary principle, or when there are actions, including advertising, that lead to the consumption of foods that are bad for health or to the distancing of traditional eating habits (BRASIL, 2013).

One of the ways to guarantee food sovereignty and sustainability from a FNS perspective is by intensifying agrarian reform and encouraging family farming, since, as mentioned above, the model of food production and consumption is fundamental to guaranteeing food security. Fome Zero (Zero Hunger), a social policy whose key element is FNS, includes incentives for family farming in its programme, and in recent years it has been successful in several respects because it has "reworked issues related to agriculture, food supply and nutrition, giving them an integrated character" (BELIK, 2012).

Public policies that support family farming are in fact encouraging economic development with greater income distribution, and are therefore combating an SDH that greatly aggravates the population's health. Promoting FNS based on the principle of food sovereignty and sustainability makes it possible to value people's culture:

> Food sovereignty is a crucial principle for guaranteeing food and nutrition security and concerns the right of peoples to define policies, with autonomy over what to produce for whom and under what conditions. Food sovereignty means guaranteeing the sovereignty of farmers, extractivists, fishermen and other groups over their culture and over the goods of nature (BRASIL, 2013).

While agribusiness, as you can clearly see in the extract from Jean Ziegler's work, leads to people's penury:

> The global empire of the agri-food trusts creates hardship and hunger for hundreds of millions of human beings - it creates death. Family food agriculture, on the other hand, provided it is supported by the state and has the necessary investment and inputs, is a guarantee of life (ZIEGLER, 2013).

In this extract, Jean Ziegler's ideas on agribusiness and family farming are related to the conditions of life and death, within the context of food security. This highlights the relationship between the different production models and their consequences for the population's health. Within the context of agribusiness, the issue of food insecurity leading to malnutrition is seen as an example of a health problem. The description below reveals the problem of malnutrition in children, which often has irreversible consequences.

> The body first depletes its sugar reserves and then its fat reserves. Children fall into a state of lethargy. They quickly lose weight. Their system
> The immune system collapses. Diarrhoea accelerates the agony. Mouth parasites and respiratory tract infections cause terrible suffering. The destruction of muscle mass begins. Children can no longer stand. Like small animals, they curl up on themselves on the floor. Their arms hang lifelessly. Their faces resemble those of the elderly. Finally, death comes (ZIEGLER, 2013, p.33).

Based on this question, we can estimate the importance of supporting family farming for FNS as one of the forms of political action to combat the social determinants of health in order to promote food security and social equity. At least in Brazil, a crucial issue has already been achieved: the understanding of the right to adequate food by the state, as the text below shows:

> The state must provide food directly to individuals or groups unable to obtain it on their own, until they are able to do so. Therefore, the obligation to provide is more particularly related to everyone's fundamental right to be free from hunger. A state must provide for the DHAA of certain individuals or groups, through cash transfers or basic income; delivery of food in accordance with the specificities of each group, population or community; or other social security schemes. (BRASIL, 2013)

Based on the notion of the right and public policy actions relating to FNS, balanced social development can be promoted. According to Becker (2010), a recent study reveals that programmes to support family farming, such as the Food Acquisition Programme (PAA), can stimulate diversified production, boost the production and consumption of ecological foods, which are foods that do not use chemical products, and sustain traditional activities. These factors include the principles of food sovereignty and sustainability, which are disregarded by the capitalist agricultural model, with implications for health promotion.

When it comes to policies for family farming, the PAA is a very interesting programme, because even though the National Programme for Strengthening Family Farming (PRONAF) generates a much larger amount of resources than the PAA, the PAA directly links family farming production to the issue of food security, since the programme values the diversity of Brazilian family farming and does not necessarily focus on production on a large scale.

The PAA has two beneficiaries: suppliers, who are family farmers, agrarian reform settlers, foresters, aquaculturists, extractivists, artisanal fishermen, indigenous people, members of rural quilombo communities, etc., and consumer beneficiaries, who are individuals in situations of food and nutritional insecurity and those served by the social assistance network and food and nutrition centres.

Therefore, in order to improve public health in terms of SA, it is necessary to act on social determinants through policies to combat health inequalities, i.e. unfair and avoidable inequalities (BUSS, 2007).

2.2 Scenario: the state of Maranhão and the discussion of agricultural education in Brazil

2.2.1 The State of Maranhão

According to data from the demographic census carried out by the Brazilian Institute of Geography and Statistics (IBGE, 2010), the state of Maranhão has a territorial area of 331,936.948 km^2 with a population of 6,574,789 inhabitants and a total of 217 municipalities. In terms of gross domestic product (GDP), it is the fourth richest state in Brazil's Northeast Region and the 16th richest state in Brazil. In spite of this, the state's HDI is 0.639, ranking it at the bottom when compared to other federal units, and the incidence of poverty, as measured by the IBGE, was 53.38 per cent (IBGE, 2013).

Figura 1- Mesoregion State of Maranhão
Source: IBGE, 2013

Maranhão currently has most of its population living in urban areas, but it is the state in Brazil with the highest percentage of its population living in rural areas. According to IBGE data in 2010, 36.9 per cent of Maranhão's 6.5 million inhabitants did not live in urban areas. This represents a universe of 2,427,640 people in the whole of Brazil.

State. As for the state's economic activity, the primary sector, agriculture and livestock, are important activities, as well as fishing. Therefore, as it is an eminently agricultural state, a large part of its population works in this economic sector in order to survive.

According to the last agricultural census in 2006, family farming in Brazil accounted for 84.4 per cent of agricultural establishments, but only 24.3 per cent of the total area occupied by these establishments. This revealed a concentrated agrarian structure. Despite this, family farming was responsible for 74.4 per cent of all jobs and makes a major contribution to the production of food for Brazilians' basic food basket, being responsible, for example, for 87 per cent of cassava production, 70 per cent of beans, 46 per cent of corn, 38 per cent of coffee, as well as supplying a large part of animal protein, such as: 58 per cent of milk, 59 per cent of pigs, 50 per cent of poultry and 30 per cent of cattle.

In summary, the data reveals the potential of this segment to provide food security and create jobs, totalling 12.3 million people employed compared to 4.2 million in non-family farming, which, discrepantly, occupies 75.7% of the total area of agricultural establishments.

Looking at this situation (land occupation and use) for Maranhão, the situation is no different. According to the same census data, family farming in the state represents 91% of agricultural establishments, thus occupying an area of 35% of the total in the state. Following the national model, Maranhão employs most of the rural labour force, 87%, with rice, corn, milk, meat

17

and poultry being the most produced foods. In terms of labour employed, family farming employs 19 people per 100 hectares, while non-family farming employs 1.6 people per 100 hectares. Therefore, in the state there is a high concentration of land owned by agribusiness, but the majority of the labour force in agricultural establishments is family farming.

Table 1: Number of agricultural establishments in Maranhão

Area Groups	1.985	1996	2.006
Less than 10 ha	445.064	272.100	136.014
10 to less than 100 ha	57.205	59.360	68.034
100 to less than 1,000ha	23.068	20.796	22.300
1,000 ha and more	2.342	1.681	1.706
no inf. or no area	3.734	14.254	58.983
Total	531.413	368.191	287.037

Source: FRANÇA; DL GROSSI; MARQUES, 2009.

Table 2: Areas of agricultural establishments in Maranhão (hectares)

Area Groups	1.985	1996	2.006
Less than 10 ha	675.994	389.795	178.368
10 to less than 100 ha	2.361.581	2.404,123	2.615,218
100 to less than 1,000ha	6.003,062	5.201,157	5.505,606
1,000 ha and more	6.507,628	4.565,617	4.692,256
Total	15.548,267	12.560,692	12.991.448

Source: FRANÇA; DL GROSSI; MARQUES, 2009.

These figures show that Maranhão follows the pattern of Brazil in terms of land concentration, as shown by Azar (2011):

Agribusiness in Maranhão, in line with the national configuration, is structured around large tracts of land, the practice of monocultures and the use of modern technological resources. These three essential elements for the accumulation of this sector imply the expulsion of peasant families, the reduction of the labour force, as well as the inadequate exploitation of natural resources, considering the need for deforestation in order to plant monocultures.

Therefore, despite the scenario that reveals a high GDP in Maranhão, with agribusiness being a strong contributor to this GDP, the advancement of a society is not revealed using only the economic dimension of development; other social, cultural and political characteristics that influence the quality of human life must also be considered. The state of Maranhão is an example of this contrast: a rich state with many poor people. One of the ways to reveal the precarious situation of a large part of Maranhão's population, especially those living in rural areas, is to find out about the food insecurity situation in which they find themselves. The prevalence of food insecure households in the state of Maranhão is 39 per cent, i.e. less than half of the population is food secure, according to an IBGE study. Of the 27 federal units, Maranhão and Piauí are the only two below 50 per cent (IBGE, 2013). In the ranking of the worst states in terms of food insecurity, Maranhão is in first place as a place where hunger is present or a cause for concern.

Figure 2: Proportion of households with moderate or severe food insecurity (%)

Source: ANDRADE, 2014.

Considering the importance of family farming in keeping people in the countryside through occupation in rural enterprises and its contribution to eradicating hunger and providing food security, public rural development policies are needed for the state of Maranhão in order to achieve equity within the state.

From the above, we realise that agriculture is the economic base of the state and that family farming is the predominant form of production, capable of generating and distributing income and guaranteeing food security for the population. With this scenario in mind, it is clear to see the importance of technical agricultural schools in this context for training workers who, with technical knowledge, will be able to contribute to the development of this state, which is predominantly rural and made up of poor small producers.

2.2.2 The discussion of agricultural education in Brazil

> Analysing the country's current economic reality and concluding that there are at least two distinct models of agricultural production, one geared towards production and consumption, identified as family farming, and the other geared towards large-scale production, i.e. commercial agriculture or agribusiness, increases the difficulties and challenges imposed on agricultural education so that it can meet the different demands, (SILVA; SILVA, 2012).

Based on the excerpt above, we can see that the author expresses concern about this form of teaching, as she believes that it must adapt to the current historical moment, regardless of the models, strategies and priorities defined, which is the challenge raised in the re-discussion of agricultural education (SILVA; SILVA, 2012). In fact, this concern was also voiced by various sectors of society that were seeking to transform this form of education.

In 2008, the Ministry of Education (MEC), through the Secretariat for Professional and Technological Education (SETEC), drew up a document together with the players involved in agricultural education that describes and systematises actions aimed at "The (Re)signification of

Agricultural Education". The proposal for this (Re)signification "originated from the need to rethink the predominant model in the institutions that work in agricultural education, taking into account the transformations in society and production processes." (MEC/SETEC, 2009)

> The document states that there is no denying that there is a strong movement in search of a sustainable production model. Agroecology, with its low input of external inputs, presents itself as an alternative that is less damaging to the environment and has a more appropriate socio-economic and financial return, capable of reducing poverty and meeting the social needs of the population. (MEC/SETEC, 2009, p.14-15)

It is clear from the document the focus given to agricultural education based on agroecology, aimed at a total change in the conception of agriculture and the world. When we seek to work with the needs of the historical moment, and this is currently revealed as thirsty for reorganisation, in other words, in search of sustainable development (economic, social and environmental), it is because we recognise that the time is ripe for rethinking, and that there is no longer any way for education to sustain an unsustainable world model.

According to the Political Letter of the Third National Meeting of Agroecology, 2014, "the agroecological perspective is also increasingly being integrated into teaching, research and extension practices through the convergent action of agroecology movements and the militancy of professionals from scientific-academic institutions." (POLITICAL LETTER, 2014)

According to Michel Foucault: "Every system of education is a political way of maintaining or modifying the appropriation of discourses, with the knowledge and powers that they bring with them". Therefore, the discussion and rethinking of this form of teaching debated here is only possible when it originates within the sphere of education itself, so the document generated by SETEC/MEC is praiseworthy, but there is a need for the political commitment of the professionals who work "at the cutting edge", as the ANA Political Charter suggests, to make it possible to realise an education that focuses on training for human emancipation, in other words, that meets the interests of society and not the market. However, it has to be considered that the issue is a little more complex, and not simply a question of political commitment, since it has to be taken into account that this professional receives from many quarters the strong discourse of agribusiness as efficient and family farming as a backward form of production, which often jeopardises their perception/action.

> In developing countries such as Brazil, there is still a great shortage of professionals who can provide technical assistance and guidance to family farmers on, above all, scientific and technological innovations in the area and better soil management, which would contribute to future generations being able to enjoy a new reality where there is rational production of food, energy and goods produced within ethical and sustainable standards. (MOREIRA, 2013, p. 53.)

Faced with this shortage (which has various causes), we realise the importance of agricultural teaching professionals committing themselves to their pedagogical practice so that it can contribute to this change. According to Guiraldelli Jr. (2006) "[...] pedagogy is the theory of education, that is, the narrative about what should happen in educational activity according to pre-established ends, according to values that one wants to preserve and reproduce and in accordance with new values that one wants to institute." Among these new values, it's worth putting into practice what the 2012 National Curriculum Guidelines for Technical Vocational Education at Secondary Level (DCNEP/2012) propose in terms of seeking socio-economic and environmental development.

In a search that was carried out by surveying the literature on research involving students on technical courses in agriculture and livestock farming, the strong environmental focus given to the research was realised, leaving aside the social focus that accompanies agricultural activities. The main focus of these studies was to understand the environmental perception of these students. The study "Environmental perception of future agricultural technicians" (BARBOSA, 2014) revealed, for example, that a large proportion of students consider the countryside to be healthier than the urban environment, revealing a strong influence of common sense and the media on people, who tend to create an image of an idealised nature when thinking about the countryside. In this sense, these students are visibly unaware of the impacts of the Brazilian rural development model and the economic, social and ecological problems it causes.

In a study on the "environmental approach in technical agricultural education curricula" (BARBOSA, 2014, p. 5), the results show that these curricula are poorly articulated with environmental education. This compromises these students' perception of the reality of the effects of land use and occupation by the agricultural sector, since environmental education would be strategic in raising awareness of the environmental damage linked to the effects of pesticide use and management.

In the search for research on the training of technicians with regard to the relationship between farming and food security (PIGATTO, 2011), one study was found that addressed this relationship by emphasising the exploitation of natural resources. In the literature review on agricultural training, no research was found that focused on teachers' and students' perceptions of the effects of agricultural activity, the way it is developed and the social determinants, particularly food security. This shows the importance of discussing and rethinking this form of teaching.

2.3 Vocational Education and Training for Agricultural Technicians

Vocational education offers technical courses, whose form of teaching was previously conceived only by teaching how to do things. Today, as a result of the changes that have taken place in the world of work, teaching has had to adjust to the current scenario, and must train technicians and professionals who can understand the relationship between science and work, acquiring a critical understanding and the social implications of doing their job (GOMES; MARTINS, 2004).

"Technological developments and social struggles have changed relations in the world of work. Due to these tensions, the existence of workers who perform only mechanical tasks is no longer acceptable. Parecer CNE/CEB n°: 11/2012, p. 6" (BRASIL, 2012). When it comes to vocational education for the training of agricultural technicians, this is any professional who has graduated from courses held at medium-level agrotechnical schools. Agricultural Technicians in their various modalities have their professional registrations with the Regional Council of Engineering, Architecture and Agronomy (CREA) in their region (BRASIL, 2014).

This agricultural technician:

> Plans, executes, monitors and supervises all phases of agricultural projects. Manages rural properties. Draws up, implements and monitors preventive sanitation programmes in animal, plant and agro-industrial production. Inspects products of plant, animal and agro-industrial origin. Carries out rural measurement, demarcation and topographical surveys. Works on technical assistance, rural extension and research programmes (BRASIL, 2012).

These professionals are trained more specifically to meet the needs of agricultural and livestock production models, whether they are made up of small and medium-sized producers or large-scale agribusiness production.

> For a long time, this absorbed a significant number of agricultural technicians, which was one of the reasons that influenced the adoption of the current model of agricultural education, orientated towards the so-called farm school, where the principle of learning by doing predominates, geared towards a large-scale conventional agricultural production system (BRASIL, 2012).

Considering the social struggles and technological developments that have changed relations in the world of work, it is essential that the training of agricultural technicians includes a dimension that involves the perception of the problems caused by forms of agricultural exploitation and their consequences for public health and the environment.

> The agricultural technician's professional work is directly related to the conservation of natural resources and, furthermore, they need to be trained in ethical values that demonstrate that the economy and the development of the agricultural sector are not above the healthy environment and the quality of life of individuals (BARBOSA, ZANON, 2010).

With this in mind, the training of professionals who will work in the agricultural sector, at a time when the country and the world are discussing sustainability in the forms of production in rural areas, makes it necessary for vocational schools to conceive that,

> As far as agribusiness is concerned, the impacts caused by the prevailing model of agriculture and the growing demands for sustainability (social, environmental and economic) and improved product quality are priorities in land use and occupation policies (GIANEZINI, 2012).

One of the evidences of the changes brought about by social struggles are the various forms of support for family farming that the current government has given. Knowing that "rural poverty, hunger in the countryside and the rural exodus are largely the result of the failure of small and medium-sized rural productive activities, especially agricultural production" (MALUF, 2013), and that the government is seeking to end poverty in the country, there is nothing more natural than acting at the source to reduce the perpetuation of this problem. In vocational education, support for family farming can be seen in the opinion drawn up by the National Education Council in 2012, which partly deals with the focus of agricultural education in vocational education.

> At the same time, they must prioritise the family farming sector and, as one of the reference elements for production dynamics, agroecology and organic production systems. In short, agricultural education institutions must pursue human development, the articulation of local groups, equity in income distribution and overcoming inequalities, and the reduction of social differences, with community participation and organisation. Similarly, issues of gender, generation, race, ethnicity, reducing poverty and exclusion, respect for human rights, reducing the environmental impacts of agricultural production and the need to improve the quality of life are also important.
> Toxic waste and pollution, the balance of ecosystems and the conservation and preservation of natural resources should be objectives to be achieved (BRASIL, 2012).

Currently, agricultural education and public policies that support family farming are considering socio-economic development from the point of view of social, economic and environmental sustainability. But it's important to know that there are challenges to be faced in order to realise these actions, given that the current economic model dictates the ways in which agricultural products are produced, marketed and consumed.

2.4 The role of the agricultural technician, his contribution to guaranteeing food security and his work in rural Maranhão.

In carrying out their work, all professionals seek to promote the development of the area in which they work, based on the knowledge they have acquired during their professional training. In the case of the agricultural technician, the fundamental role to be developed must be that of rural development agent through mediation. "Development agents can be represented by agricultural technicians, extension workers, agronomists, veterinarians, health workers, nurses, literacy workers, social workers and educators" (DEPONTI; ALMEIDA, 2008, p. 35).

> Through development agents or mediators, two worlds, or rather two worldviews, two systems of meaning, come into interaction. The exercise of mediation consists of the dissemination and construction of knowledge, behaviours, ideas and values that will be transmitted and will lead to the incorporation of new behaviours, identities and worldviews (DEPONTI; ALMEIDA, 2008, p. 35).

It is up to the professional educational institution to provide technicians with training that enables them to act consciously and critically on the importance of their mediation in rural areas, regardless of the agricultural production model in which they may be employed in the future. Hence the importance of the role of teachers in training these technicians.

> Therefore, in order to train professionals who are aware of the systemic vision of the area in

which they work, teachers must emphasise the intrinsic ethical values of the profession, the attitude of citizenship when exercising their profession, especially with regard to the preservation of health, the environment and quality of life, making students co-responsible for building their competence (GOMES ; MARTINS, 2004).

With regard to the professional practice of agricultural technicians in Maranhão in relation to the preservation of health, the environment and quality of life (which implies food security), it is important that they are aware of the state's reality in order to know how public policies in general can contribute to ensuring food security for the population of Maranhão.

Maranhão's widespread poverty is reflected in its social exclusion indices. Of the 100 municipalities in Brazil with the highest rates of social exclusion, 35 are in this state, leading to the need to promote policies aimed at development. However, actions in Maranhão can be analysed more as structuring for big business than as development or reducing social exclusion (RODRIGUES; ALENCAR, 2011).

Knowing the real situation in the state of Maranhão, which is marked by poverty and social exclusion, makes it relatively easy to understand the conditions surrounding the health and quality of life of the population living in the state. It is difficult to accept that there is a lot of investment in the state related to the agricultural and mining sectors, while social inequalities remain, "fuelling" the poverty that generates food insecurity.

This inequality is reflected in the intensity with which capital manifests itself in the territory. This, in turn, has serious consequences for the local population, from unemployment to infrastructure conditions. To illustrate the ways in which capital manifests itself in the territory of Maranhão in an unequal and combined manner, we take as our starting point the large mining and metallurgical projects, as well as the large-scale agricultural projects. These are one-off investments that concentrate a large part of the state's production, generate few jobs and even dismantle what existed before, with the production of babassu, rice and peasant agriculture (RIBEIRO JUNIOR;CASTRO JÚNIOR, 2012).

Maranhão is currently enjoying a boom in soya production in the south of the state, especially in the region around the municipality of Balsas. Agribusiness has gained prominence and a lot of encouragement from the state government. One of the consequences of the incentives for large-scale agriculture is the decrease in the rural population due to the surplus of labour in the countryside and their departure for the big cities.

According to official data, in 1970 the rural population of Balsas was 63.4%, in 1980 it was 42.8%, in 2000 it was only 16.7%, and in 2010 it was 12.8%. Although this is a general trend, in the municipality of Balsas there has been a more marked reduction in the rural population, as agribusiness cannot develop to its full potential without breaking the ties of at least part of peasant agriculture (RIBEIRO JÚNIOR; CASTRO JÚNIOR, 2012).

Considering the reality of rural Maranhão, we can see the importance of family farming as a way of promoting more equitable rural development, guaranteeing food security for many people. The role of the agricultural technician as a mediator of rural development that seeks socio-economic and environmental sustainability is therefore fundamental, obviously through support for family farming.

Family farming is of fundamental importance in the development process in its various dimensions. From the point of view of guaranteeing the survival of families, it can be seen that production for consumption has guaranteed food security for countless families living in rural areas. As well as self-consumption, it has the capacity to supply volumes of food to the market and enable access to food. Studies have refuted the view of the 1970s that large agribusinesses were responsible for the supply of food, dismissing family farming (FAVACHO; SOUSA, 2006).

In addition to the problem of food insecurity as a result of the social exclusion caused by agribusiness in the state, another very serious problem that agribusiness generates is the great exploitation of natural resources, such as the exploitation of water, soil and the great contamination by the use of pesticides resulting from monoculture.

Just as the Green Revolution brought technological innovation and agricultural education had to train professionals to meet the production models once considered the salvation of food production, today, with the changes brought about by social struggles and new technologies, vocational education must, as mentioned before, promote in these students "an attitude of citizenship when exercising their profession, especially in terms of preserving health, the environment and quality of life" (GOMES; MARTINS, 2004, p.45).

IFMA's official documents state that its mission is to promote teaching, research, innovation and extension, with the aim of training critical, ethical and responsible citizens with a holistic and entrepreneurial vision, capable of developing sustainable actions in order to meet society's needs (MARANHÃO, 2011).

Therefore, as well as being able to act as a mediator contributing to the dissemination of knowledge and values for rural development, the role of the IFMA-trained agricultural technician in social policies, in terms of dealing with local problems and their contribution in terms of guaranteeing food security, can and should be, based on research, innovation and extension, to work towards the development of sustainable actions and to meet the needs of the population of Maranhão, which has historically experienced situations of exploitation and expropriation in the countryside, generating the poverty seen today.

Thinking about the real possibilities of working in the agricultural labour market in rural Maranhão, it can be seen that the greatest opportunities for agricultural technicians to work professionally are as extension workers and not as employees of companies linked to agribusiness. It can be seen that most of the agricultural technicians who have been placed on the job market have been absorbed by teams working on programmes linked to family farming (Pronaf, PAA and others) and in municipal agriculture departments, whose main audience is the settlements. Thus, the extension market has generated the main jobs for agricultural technicians and agronomists in the

Maranhão countryside, despite the fact that many students during their training period envisage a salaried job in large export-orientated companies.

It should be noted that the training of professionals who will enter the labour market is faced with a major challenge: the knowledge acquired during their training process shapes their way of perceiving, leading students to be more or less capable of perceiving the social and environmental implications of agricultural activities. Certainly, the way an individual perceives determines the way they react, in this case, the way they act professionally. This should be committed to economic and social development, taking into account ethical, cultural and environmental aspects.

With regard to the idea of *perception,* Fritjof Capra (1996) in his book *The Web of Life,* states that the main social and environmental problems of our time should not be understood in isolation, as they are systemic problems, i.e. they are interconnected. These problems, in fact, are facets of a single crisis, the crisis of perception, characterised by the inadequate perception of reality by most of us, evidenced in the practices of major social institutions. To quote Capra:

> Recognising that a profound change in perception and thinking is needed to guarantee our survival has not yet reached the majority of leaders in our corporations, nor the administrators and professors at our great universities (CAPRA, 1996, p. 24).

CHAPTER 3

BACKGROUND

The Food and Agriculture Organisation of the United Nations (FAO) celebrated 2014 as the International Year of Family Farming (IYFF). The aim was/is to reposition family farming at the centre of agricultural, environmental and social policies in order to promote equitable and balanced development, as well as to raise the profile of family farming and small farmers, focusing the world's attention on their important role in eradicating poverty, hunger, providing food and nutritional security, protecting the environment and others.

During the course of this study, we sought to establish and develop the debate on the relationship between land use and food security. This debate is considered extremely significant in Brazil, a country that has an immense area of land used for agricultural production, is rich in food production, but has a large proportion of its population living in poverty and food insecurity.

In this way, we can include among the factors with an impact on health related to agricultural production activity, the expansion of agribusiness as a factor in the decline of family farming and other productive activities of small landowners as a determining factor in the displacement of the population to the outskirts of urban centres and the consequences of this rural-urban migratory flow (precarious housing conditions, lack of access to basic sanitation, insufficient provision of education and health services, exposure to various forms of violence - all common problems on the outskirts of large Brazilian cities).

In Maranhão in particular, agribusiness has been growing and gaining economic prominence with soya production in the centre-south of the state. This is a production practice that leads to the exhaustion of natural resources, especially water and soil, as well as causing other problems, such as the intensive use of pesticides and changes to labour patterns in the countryside.

Based on a basic understanding of the social and environmental effects of agricultural activities and the commitment to realising an education geared towards sustainability and improving the population's quality of life, it was necessary to find out how vocational education in agriculture relates the themes of food production, product distribution and sustainability, in order to understand their consequences for health, especially with regard to food safety.

If this education does not include a critical perception, it means that this form of teaching is possibly moulding future professionals to the logic of the job offer and the market, going against the world view that seeks a full education focused on training professionals who can understand the relationship between science, society and working practices, acquiring a critical understanding of the social implications of their professional activity.

I believe that by studying the relationship between small-scale farming in rural areas, social indicators and their repercussions on human health, we can find important aspects that are not covered in the training of agricultural technicians. By identifying this gap, it would be possible to propose not only measures linked to teaching, but also new research to assess the importance of raising awareness about the role of family farming in guaranteeing food security for certain population groups, among other relevant aspects.

CHAPTER 4

OBJECTIVES

4.1 General

To find out the perception of students and teachers on the vocational training course for agricultural technicians at IFMA-Maracanã regarding the promotion of knowledge and reflection on the relationship between the way land is occupied and used (expansion of agribusiness, intensive use of pesticides and incentives for family farming) and the population's food security.

4.2 Specific

a) To identify whether and how students and teachers at IFMA - Campus Maracanã relate the growth of agribusiness to the decline in land use for family farming;

b) To identify whether and how students and teachers at IFMA - Campus Maracanã relate the decline in family farming to the state of food security of the population;

c) To find out about the curricular content of the technical course in Agriculture at IFMA - Campus Maracanã, relating it to the perception of students in the final period of the course on the themes of economic development, the expansion of agribusiness, family farming, production and access to food, food security and health.

CHAPTER 5

METHODOLOGY

5.1 Delineation

As for the design of this study, it is an exploratory study in which qualitative research methods were applied to assess knowledge about social inequalities and food security (with an emphasis on the characteristics of food production by family farming and agribusiness) in São Luís (MA). The unit of analysis studied was a federal agricultural school, the oldest in the state of Maranhão.

5.2 Context of the study

The school selected for the research is located in the municipality of São Luís in the state of Maranhão, an eminently agricultural state. According to IBGE data, in 2010 the population of the municipality of São Luís was estimated at 1,014,837 inhabitants, with an area of 834.785 km^2 , a Gross Domestic Product (GDP) of R$ 17.9 billion, a Human Development Index (HDI) of 0.768 (UNDP, 2010) and its main economic activities being industry, commerce and services. (IBGE, 2010).

> The reworked and redefined subsequent technical course in agriculture is in line with Maranhão's activities in the primary sector through the production chains of grains, fruit growing, livestock (dairy, feed, meat production/processing, leather and footwear), forestry, cotton, fertiliser production, aquaculture and mariculture (COURSE PLAN, 2009).

It is clear from the state's reality how important the institute is in articulating the socio-economic development of the region, prioritising the production chain of the state to which the course plan is linked.

IFMA-Campus Maracanã is one of the oldest institutions in the state of Maranhão specialising in secondary agricultural vocational training and is a benchmark for vocational education in the region. The campus reaffirms its commitment to the teaching-learning process in terms of vocational training aimed not only at doing, but also at students entering the labour market knowing how to do and why to do, emphasising and valuing human beings as citizens who are aware of their place in society (COURSE PLAN, 2009).

Figure 03 - IFMA-Campus Maracanã
Source: Google, Maps (2015).

The Maracanã Campus is one of 26 units of the Federal Institute of Science and Technology in the state of Maranhão. It was elevated to the status of a federal autarchy in 2008, when the former Colégio Agrícola or Escola Agrotécnica de São Luiz, now 67 years old in the state, was included in the network of federal technological schools.

Figure 04 - Location of IFMA/Campus Maracanã

Source: Google, Maps (2015).

The IFMA Maracanã Campus currently offers several courses, and the Subsequent Agricultural Technician Course is just one of the basic education modalities offered by the Institute.

5.3 Qualitative research

Using qualitative research methods, the knowledge, practices and perceptions of students and teachers at IFMA-Campus Maracanã were assessed.

When it comes to qualitative research, "it works with values, beliefs, representations, habits, attitudes and opinions and is suited to delving into the complexity of particular and specific phenomena, facts and processes of groups that are more or less delimited in size and capable of being covered intensively" (MINAYO; SANCHES, 1993, p. 239). However, it has some limitations, such as its results, which are not reproducible, and the subjectivity of the researcher involved in the research itself, which can intervene in the analyses.

The book "The Art of Research" by Goldenberg (2004) reveals from its very title the researcher's capacity for the creativity required in the process of carrying out research that reveals the truest possible results. The author begins the book by talking about positivism, which considers that all sciences should be objective and neutral so that they result in generalisations without the intervention of the researcher's subjectivity.

However, it also reveals that the social sciences, whose object of research is emotions, beliefs and values, cannot be treated as such, since in order to understand the object, it is necessary to interpret the experiences of the person being researched, thus revealing a relationship between the object and the researcher. This relationship is necessary to understand the complexity of the social phenomenon as seen through the eyes of others. The author defends the qualitative method for the social sciences, as the numerical representativeness of the quantitative method limits the understanding of the reality studied (GOLDENBERG, 2004).

According to Minayo (2001): "one of the questions posed about the scientificity of the social sciences concerns the plausibility of dealing with a reality in which both researchers and investigated are agents: wouldn't this order of knowledge radically escape all possibility of objectification?". Considering the risk of analysing qualitative material due to the possibility of the researcher interfering with their subjectivity, there is "a need for a methodological effort that guarantees objectification, that is, the production of an analysis that is as systematic and in-depth as possible and that minimises the incursions of subjectivism, guesswork and spontaneity". (MINAYO, 2001)

In this way, using a qualitative approach, we sought to understand the opinions, representations and values expressed by the students after receiving their technical agricultural training, the lessons they received and how they relate to the social context of the agricultural labour field.

5.3.1 Research Subjects

The research subjects were students and teachers from the Federal Institute of Education, Science and Technology of Maranhão - Campus Maracanã, located at Av. dos Curiós s/n, Vila Esperança, São Luís (MA).

The size of the sample was based not on quantity, because "the representativeness of data in qualitative research in the social sciences is related to its capacity to make it possible to understand the meaning and the 'dense description' of the phenomena studied in their contexts and not to its numerical expressiveness. " (GOLDENBERG, 2004).

In this case, a minimum sample size of ten students and seven teachers was set, given that in qualitative research some authors suggest a minimum of six participants and a maximum of twelve for focus groups. It should be noted here that the teachers interviewed taught more than one subject in the curriculum, which increases representativeness.

a) Inclusion criteria

The study included students from the last period of the course, of both sexes, from the IFMA's technical course in agriculture in the subsequent modality.

The teachers included in the study were men and women from the IFMA's technical course in agriculture, in the subsequent modality, who taught technical subjects to the students who were the subjects of the research.

b) Exclusion criteria

There are no exclusion criteria

5.3.2 Data collection

Firstly, there was a survey (interviews) involving teachers and students (last term) from the technical course in agriculture at the Federal Institute of Maranhão (IFMA), Maracanã Campus.

The second stage involved a survey of the training policy for Agricultural Technicians at IFMA, the curriculum matrix and content.

5.3.2.1 Semi-structured interviews

Focus groups and individual interviews were carried out, both with all the participants in the research, since we aimed to study both what is offered by the IFMA in the training of students and what teachers think about issues such as economic policy and the development model adopted, the expansion of agribusiness and its environmental and social repercussions, the migratory flows triggered by the lack of incentives for family farming and the occupation of the countryside by monoculture, the relationship between food security and health, as well as issues such as citizenship, inequality, etc.

To collect data in this phase, two focus groups were held, one with students and the other with teachers, as well as individual interviews using a semi-structured questionnaire and an audio recorder.

a) Focus groups and individual interviews

- Iª PHASE: presentation of the project and recruitment of students and teachers

After this project was approved by the Research Ethics Committee of the National School of Public Health (CEP/ENSP), I introduced myself to the school's teaching department and to the potential participants in the research. With regard to the students, a visit was made to their classrooms, and with regard to the teachers, a visit was made to the teaching department, the sector to which the teachers on this course are linked.

It was explained that the aim of this study was to find out how future agricultural technician professionals were trained, taking into account their knowledge of the relationship between social inequalities and food security (with an emphasis on the characteristics of food production by family farming and agribusiness). After the presentation, people who wanted to take part in the research as volunteers were asked for their names and contact details, and participants were recruited so that a date could be set for the focus group and interview. The interviews and focus groups lasted an average of 30 minutes for the individual interviews and an hour and a half for the focus group.

The participants' consent to take part in the interviews and the focus group was requested and formalised by reading and signing the Informed Consent Form (ICF), one copy of which remained with the interviewee and the other with the researcher. Everyone was free to participate and withdraw.

• 2ª PHASE: Individual interviews and focus groups with students and teachers

In the second phase, the aim was to highlight the key elements of the students' and teachers' training, in order to find out whether their training actually included knowledge that promoted a perception of the relationship between the way land is occupied and used and the population's food security. To this end, individual semi-open interviews were held with eight students and five teachers, and topics were chosen for discussion in each focus group, with nine students and six teachers who were part of the plant and animal production axis. The focus group was held first, followed by the individual interviews.

The questions to the students covered the relationship between the growth of agribusiness and the decline in land use for family farming, food production and access, food security and health. The main topics covered in these interviews are presented in Appendix 01.

The teachers were asked the same questions as the students, but aspects related to their work experience were added. These elements were related to their perceptions of agribusiness, family farming, food production and their effects on food security. The main topics covered in these interviews are presented in Appendix 02.

The focus group data was collected by choosing a topic to discuss (APPENDIX 03 AND APPENDIX 04), which allowed us to get to know the students' and teachers' perception of the issue involved. The location for the interviews and focus groups was a classroom and/or teacher's room, decided by the school's Education Department and the teachers. The audios were transcribed immediately after the interviews.

The Principal's consent was obtained for the data to be collected at the school (APPENDIX 01) and approval by the Research Ethics Committee of the National School of Public Health (ENSP/FIOCRUZ) was obtained through opinion No. 749.482 of 27 September 2014. This research followed the current guidelines and regulatory standards for research involving human beings, Resolution 466 of 12/12/12.

5.3.1.1 Survey of didactic content

At this stage, documents were collected on:

a) The National Curriculum Guidelines for Professional Technical Education, currently in force in Brazil;
b) Document on the National Curriculum Guidelines for Professional Technical Education that governed the running of the course in progress;
c) Of the course plan for the subsequent technical course in agriculture on this campus.

The purpose of this survey was to find out about the training policy for agricultural technicians, with an emphasis on the curriculum, subjects and general content of the agricultural technician training course at IFMA.

5.3.2 Data Analysis

5.3.2.1 Discourse of the collective subject

Many studies show that qualitative methods provide very significant and dense data, but are also very difficult to analyse. In this study, the Collective Subject Discourse (CSD) technique was used to tabulate and organise the data, according to the methodology proposed in the late 1990s by Lefèvre and Lefevre (2005). The Discourse of the Collective Subject has its theoretical foundations in Social Representations.

According to Lefevre and Lefevre (2005), the application of the CSD technique to a large number of empirical studies has demonstrated its effectiveness in processing and expressing collective opinions. The proposal of the DCS as a qualitative research technique makes it possible to retrieve the wide range of meanings of social representations that are included in the culture of a society. "The use of CSD is being widely disseminated in scientific research, bringing a significant change in quality and efficiency, revealing in detail the representations, beliefs, values and opinions on a specific subject." (ALVÂNTARA, VESCE, 2008)

According to Lefevre and Lefèvre (2003) "the Collective Subject Discourse or CSD is a synthesised discourse made up of pieces of discourse with a similar meaning brought together in a single discourse. Based on the theory of Social Representation and its sociological assumptions, the CSD is a technique for tabulating and organising qualitative data that solves one of the great impasses of qualitative research in that it allows, through systematic and standardised procedures, to aggregate statements without reducing them to quantities."

> The Discourse of the Collective Subject is therefore an explicit proposal to reconstitute a collective empirical being or entity, expressing an opinion in the form of a discourse subject in the first person singular. These discourses reveal social representations based on the socio-cultural formation of a group (LEFÈVRE; LEFÈVRE, 2006).

5.3.2.2 The collective subject discourse technique

For the CSD technique proposed by Lefevre and Lefevre (2003), the meaning present in collective opinions is recovered to form a set of collective discourses or CSDs. For this process of producing CSDs to take place, four operators/operations are required:

    1. key expressions (E-CH)

    1.1 Central Centres (CIs),

    3. Anchorages (ACs),

    4. Collective Subject Discourses (CSDs) themselves.

This series of operations is carried out on the verbal material collected in the research and each one has a function. The E-Ch are selected excerpts from each statement that best describe its content. The CIs in the CSD technique are the equivalent of the Categories and describe the meaning(s) (values, ideologies, beliefs) present in the statements of each answer and also in the sets of answers from different individuals, which present similar or complementary meanings, revealing what people think.

The CA is a statement that contains a value, a theory, an ideology, a belief made explicit in the subject discourse, appearing in concrete form, but recording the CA is not an obligatory step in the CSD methodology. The DCSs are the gathering of the E-Chs present in the statements, which have CIs and/or CAs of similar or complementary meaning.

Following the steps proposed by Lefevre and Lefevre (2003), the interviews were listened to and transcribed verbatim for a better understanding of each subject's statements. Based on the transcribed verbal material, the CIs and E-Chs in each CI were identified. After identifying the CIs, the CIs with equivalent or complementary meanings were grouped together, creating a synthesised CI that best expressed all the CIs with the same meaning, equivalent meaning or complementary meaning.

To construct the DCS, all the E-Chs in the same grouping were copied, without giving up the rules of having a beginning, middle and end, going from the most general to the least general and most particular and written in italics.

Below is an example of the construction of the DCS and the identification of the E- Ch and IC from the statements of eight subjects to the question below.

1. The year 2014 was declared by the FAO as the international year of family farming. Was this discussed at any point during the course? In what way?

IAD Discourse Analysis Tool 1

	Key Expressions (ECH)	Central Ideas (CIs)	Anchoring (AC)
01	Tiago (Iª Ideia) ...There was no such discussion here on the course...	(Iª Idea) There was no such discussion A	
02	John (Iª Idea) ...I don't remember...	(Iª Idea) I don't remember if there was this discussion B	
03	Maria (Iª Idea) As far as I remember, there was nothing about this in our course...	(Iª Idea) Not that I can remember B	
04	(Iª Idea) No	(Iª Idea) There was no such discussion A	
05	(Iª Idea) No	(Iª Idea) There was no such discussion A	
06	Matthew (Iª Idea) No	(Iª Idea) There was no such discussion A	
07	Daniel (Iª Idea) No	(Iª Idea) There was no such discussion	

08	Sara.(Iᵃ Idea) No	A (Iᵃ Idea) There was no such discussion A	

Discourse Analysis Tools IAD 2

A. The student says there was no such discussion.

Key Expressions	DSC
Tiago (Iᵃ Ideia) ...There was no such discussion here on the course... (Iᵃ Idea) No (Iᵃ Idea) No Matthew (Iᵃ Idea) No Daniel (Iᵃ Idea) No Sara.(Iᵃ Idea) No	*There was no such discussion here on the course*

B. The student doesn't remember if this discussion took place.

Key Expressions	DSC
João (Iᵃ Idea)...I don't remember... Maria (Iᵃ Idea) As far as I remember, there was nothing about this in our course...	*As far as I can remember, there was no mention of this in our course*

5.3.3.3 Analysing didactic content

We read and analysed the didactic content of the technical course in Agriculture at IFMA - Campus Maracanã, in order to relate it to the perception of the students of the last period of that course on the themes of economic development, the expansion of agribusiness, family farming, production and access to food, food security and health.

5.4 Ethical aspects

The questionnaires and interviews were carried out with the prior signature of the informed consent form (APPENDICES 5 and 6). All the participants were informed about the objectives of the research, the confidentiality of their identities with regard to the results found and that their refusal to take part would not harm them in any way.

The interviewee's privacy has been respected and their name will never be identified in any reports or publications that may result from this research. Ownership of the information generated will be for the exclusive use of the researcher responsible, ensuring that no-one has access to the data, in order to preserve the confidentiality of the information, in accordance with Resolution 466/12.

The research project was submitted to FIOCRUZ's ENSP and approved on 27/09/2014 (CEP/ENSP research protocol No. 749.482, dated 13/08/2014 - CAAE: 33635414.8.0000.5240).

CHAPTER 6

RESULTS AND DISCUSSION

6.1 Descriptive data on the research subjects

The individual interviews were made up of five teachers and the focus group of six teachers of technical subjects in the specific area of the subsequent technical course in agriculture at IFMA-Campus Maracanã who taught subjects in the classes of the students taking part in this research. As for their professional backgrounds, the group was made up of two graduates in Agricultural Sciences and four Agricultural Engineers.

As for the students, eight people were interviewed and the focus group was made up of nine people in the final stages of completing the course.

6.2 Analysis and discussion of the teachers' statements in the interviews

The results of the studies were presented in the form of the discourse of the DCS constructed from the answers to the questions in the individual and group interviews addressed to the students and teachers taking part in the research. Analyses of the central ideas (CIs) and the CSDs enabled them to be grouped into six themes:

1. The FAO declared 2014 the International Year of Family Farming;
2. Food production for agribusiness and food production by family farming.
3. Agribusiness, family farming and labour in the countryside.
4. Large estates and small properties - social and economic repercussions.
5. The right to food security. Food production and distribution. Inequities, scarcity and hunger.
6. Food safety and health

1 . The FAO declared 2014 the International Year of Family Farming.

Two central ideas emerged from this first theme. When teachers were asked if at any point during the course they had commented on or discussed 2014 being declared the International Year of Family Farming, four teachers formed DCS 1, with only one teacher forming DCS 2.

Central Idea 1: The teacher didn't work on the theme of the International Year of Family Farming during the course

DSC1: *During the course I didn't comment on 2014 being celebrated as the international year of family farming. In fact, I wasn't even aware of it.*

Central Idea 2: The teacher worked on the theme of the International Year of

Family Farming

during the course with a focus on Agrarian Reform

DSC2: *The focus given to this question was on land reform in the country, which has not really benefited all the necessary aspects that family farming needs, especially in terms of infrastructure.*

The idea expressed in the speech reveals the lack of knowledge on the part of agricultural professionals about a celebration that took place at international level and that is important in their area of professional activity. It is clear that there should have been greater involvement on the part of the teachers and the institution, given that at that time it had the mission of "Promoting teaching, research, innovation and extension, with the aim of training critical, ethical, responsible citizens, with a holistic and entrepreneurial vision, capable of developing sustainable actions in order to meet the needs of society" (IFMA, 2011).

Thus, since family farming is recognised as a form of production that plays an important role in eradicating poverty, hunger, providing food and nutritional security and protecting the environment (FAO, 2013), it can be considered that a good opportunity was missed to involve students in a discussion in line with the institutional mission.

Considering that this group of professionals works in a teaching, research and extension institution with a focus on agricultural education and that they are possibly involved in scientific research, research projects and taking part in meetings in the field, the discourse shows the low reception and link with the theme of family farming at a historic moment of celebration.

In CSD 2, although the focus of the question was on agrarian reform (an approach that seems appropriate to the Brazilian reality), the number of teachers who worked on the subject (only one) showed the weakness of the discussion suggested by the FAO. This weakness is evident when none of the students interviewed

confirmed that he had been working on the issue of family farming since the UN celebration in 2014.

Possibly, if the subject were included as content or a theme to be worked on in class in the most diverse subjects, students would be able to assimilate the subject in greater depth, which would help them to recognise aspects relating to family farming as important to their technical training process.

2 . Food production by agribusiness and family farming

Six central ideas emerged from this second theme. When the teacher was asked if he had dealt with the differences in the way food is produced by agribusiness and family farming and what kind of reflection he would have led the student to have on the subject, three teachers said they had dealt with these differences and two said they had not.

Central Idea 1: Yes, in the course the teacher talked about the differences in the way
food is produced by agribusiness and family farming.

> CSD 3: *I discussed the issue of production without the use of agrochemicals. I also focused on the Pronaf project and projects with other resources of my own, emphasising that family producers don't have access to credit lines, unlike agribusiness. But I focused mainly on agribusiness, on commodities in terms of numbers, emphasising how Brazil is placed in the world. However, I also always talked about the issue of small rural modules, small producers, who are able to work in small areas, but using advanced technologies.*

Central Idea 2: No, the teacher didn't talk about the differences in the way food
is produced by agribusiness and family farming.

> CSD 4: *I didn't discuss the difference between the forms of food production. I'm against this distinction between agribusiness and family farming, because I see it as the same thing, because family farming doesn't donate the food, so everything you sell that comes from agricultural work is agribusiness.*

Those who worked on these different forms of production focused on them according to the subject they taught, which seems natural. The focuses chosen to show this distinction were: production without the use of agrochemicals, agribusiness and commodity production, small producers, as well as the differences in access to credit for each form of production. In this way, students can understand these distinctions from various angles, giving them a broader perception of each form of production.

It's worth noting that the teacher highlighted the different way in which the government treats financing for family farming and agribusiness, when he says *"family farmers don't have access to credit lines, unlike agribusiness"*. The accusation in the teacher's speech regarding the privileging of agribusiness (when it comes to funding) allows the student to realise that there is a shortage of rural development policies aimed at small producers, those who generally produce without the intensive use of agrochemicals (a topic discussed in class) *"I discussed the issue of production without the use of agrochemicals a lot"*.

This can lead them to the following line of thought: those who have more sustainable forms of production in Brazil have less investment from the government. This perception provides a critical perception, allowing students to know how policies aimed at that sector operate or the focus of policies aimed at that sector, thus meeting the objective proposed by the course plan. "To provide technical scientific knowledge within a critical-reflexive attitude of citizenship in the exploitation of agricultural areas" (COURSE PLAN, 2009)

ppoémn always also talking about the issue of small rural modules, of small producers," in this excerpt from the speech we can see that the different dynamics of these forms of production are recognised and taught in the classroom, emphasising the size of the properties. *"I focused the project on the Pronaf line."* This approach allows students to understand the need for these small producers to participate in specific funding programmes, thus learning about the National Programme for Strengthening Family Farming (Pronaf), which finances individual or collective projects that generate income for family farmers and land reform settlers. (BRASIL, 2015)

As for the teachers who didn't work on this distinction in class, either because they chose not to take into account important differences between these forms of production, *"because I understand it as the same thing, because family farming doesn't donate the food, so everything you sell from an agricultural sector is agribusiness",* or for another reason, they end up not addressing, in addition to characteristics linked to the economic dimension, those that we can include among the social and environmental effects of each of them.

Pedagogical choices like these can jeopardise students' view of the value or importance of each form of food production for society, especially in terms of food security, because when they are not presented with the distinction between them, preserving a space for reflection, discussion and criticism about, for example, the environmental and social problems that each form of production brings with it, students tend to perceive as the best form of production the one that generates the most profit, because the "business" value becomes the only weight of the issue. And the role of family farming in food security and in reducing hunger and poverty is not given the weight it is recognised as deserving.

According to the FAO, it is understood that:

> Family farming includes all family-based agricultural activities and is linked to various areas of rural development. Family farming consists of a means of organising agricultural, forestry, fishing, pastoral and aquaculture production that is managed and operated by a family and is predominantly dependent on family labour, both women and men (FAO, 2014).

Brazil's understanding of what family farming is in Law 11.326 of 24 July 2006 (the Family Farming Law) is in line with the FAO's understanding. While the understanding of agribusiness is:

> The current use of the term "agribusiness" in Brazil expresses - or wishes to express - agricultural activities that use intensive production techniques (mechanisation and chemistry) and scale, which generates an increase in production and productivity (SAUER,; LEITE, 2008).

Considering the important role of the teacher in educating students to understand not only the difference between these forms of production, but also how the transition from traditional agriculture to capitalist agriculture came about or the reasons for it, it is necessary for teachers to strengthen their understanding of concepts that have already been established (including by law) in their area of work, as it seems that the meanings of family farming and agribusiness were not made sufficiently clear to this group of teachers during their training process or when passing them on to their students.

When asked about the environmental advantages of food production by family farming, three teachers said they had addressed the issue and two said they had not.

Central Idea 1: The teacher discussed the environmental advantages of
family farming for food production

.

> DSC5 *One of the advantages I talk about in class is the integrated farming system - crops, livestock and forestry - so that small producers can work in the same area without having to move, making the soil live and producing for many, many years, unlike the monoculture farming practised today. But I don't always talk specifically about family farming, I approach the subject from the point of view of safety and the issue of respecting certain rules and concepts that exist regarding the production of food in a sustainable way, with a view to sustainability, and it's not just a question of family farming.*

Central Idea 2: The teacher didn't address the environmental advantages of
family farming for food production.

> DSC6 *No, not that. For example, as I've already mentioned and I'll repeat, family farming is not a saint, it's in the capital market and the first aggression it causes is not wanting to use any technology... it wastes water, it doesn't have a professional, it doesn't use meteorological data, it doesn't use any research for its production, so if you put it on the scale, it's doing more harm than what you call agribusiness.*

In discourse (DCS 5), as advantages for the environment, *the system of integrated cultivation, crops, livestock and forestry"* was explained in class, showing how *the* concepts learnt in class can be applied in practice, *"some of the concepts that exist with regard to producing food in a sustainable way"*. While working on sustainability concepts, the student realises how this can happen in practice. In this way, students can see that it is possible to produce food in a diversified way and with less environmental impact.

The group of teachers who worked in the classroom with this thought are in tune with Sachs' (2012) thinking when he argues that *"we must limit the devastating impact of extensive livestock farming on forests by learning to raise cattle in a way that is more integrated with small-scale family farming"*. However, even in this same speech, it is revealed between the lines that the discussion on sustainability takes place in a general way, not always relating the role of family farming to environmental protection. Thus, the environmental discussion seems to take place without properly characterising the differences in the impact of different agricultural practices on the environment and human health

In this way, the course plan's objective of finding ways to promote sustainable progress - "Keeping abreast of production market trends in the agricultural field, valuing sustainable development as a priority goal for achieving progress" - seems not to be adequately valued in the training of agricultural technicians, since little was mentioned about the relationship between the environment, sustainable development and family farming.

From the themes of "protecting the environment" and "the practice of family farming," to a discourse such as (DSC6) *"if you put it in the balance, it (family farming) is doing more harm than what you call agribusiness"*, there is a concern about the teacher's perception of the environmental impacts specific to each form of production and the way in which this is constructed in these circumstances, the student's understanding of what safe food production actually is and what sustainability means.

On the basis of the speeches analysed, it is clear that the principles of environmental preservation are not very clearly addressed and that there is little emphasis on aspects that value the alignment between family farming and the fundamental concepts of environmental protection and sustainability.

When asked about the environmental impacts caused by agribusiness in their classes, four teachers tackled the subject and only one did not. The discourse (DCS 7) showed that this topic was dealt with on several fronts: monoculture, mechanisation, slurry, excessive use of pesticides.

Central Idea 1: The teacher addressed the environmental impacts of
agribusiness

DSC7 *Yes, I'm talking about impact in general, not just agribusiness, but impact in general, because my subject is a technical subject, aimed at technical education. I dealt mainly with the practice of monoculture in rich regions of the country, such as the southeast, which has arable land and is losing fertility due to the planting of sugar cane and soya. I also talked about the excessive use of agrochemicals, mechanisation and slurry.*

Central Idea 2: The teacher didn't address the environmental impacts of
agribusiness

DSC8 *Agribusiness is hard work ...Business is bad for those who don't want to work in the rural sector...I approach it by showing where there are job opportunities for those who want to improve their lives and work, roll up their sleeves and go out into the sun and earn money, an approach*

In fact, many of the impacts of this branch of production were dealt with, but once again the discourse reveals *THE* superficiality of the discussion: "Yes, I *talk about impact in general, not just agribusiness,"* showing that the topic is dealt with, but the discourse reveals that the teacher doesn't feel obliged to deal with it in a more academic way, with greater scientific rigour because it's *a* technical course: *"because my subject is a technical subject, aimed at technical education".* The discourse reveals the teacher's lack of commitment to presenting the various dimensions involved in this discussion in the classroom.

In this way, there seems to be an assumption that technical training lacks a systemic vision, contextualising the student's future activities and the area in which they will work. According to Gomes and Martins, the teacher must emphasise values, especially in terms of preserving health, the environment and quality of life, making the student co-responsible for building their competence (GOMES ; MARTINS, 2004).

3 Agribusiness, family farming and labour in the countryside

Two central ideas emerged from this second theme. When asked whether the course had addressed the question of which sector absorbs more rural labour, agribusiness or family farming, three teachers said it had and two said it had not.

Central Idea 1: During the course, the teacher said which sector absorbs the most labour in the countryside

> DSC9 *I took into account the technologies used in productivity to meet the needs of the amount of people there are to feed, but I commented that when you use technology you exclude labour, but productivity increases. Family farming, on the other hand, absorbs a lot of labour, not just your own family members, but also specialised labour, such as a technician or agronomist. To make the point, I mentioned that per hectare, family farming employs 10 people, while agribusiness only employs one.*

Central Idea 2: The teacher didn't talk during the course about which sector absorbs the most labour from the countryside

> DCS10 *I didn't discuss which sector absorbs more labour, because as I've already explained, I don't work with that division. As I see it, "Mrs Maria" is doing agribusiness by selling lettuce.*

According to IBGE data, family farming employs 15.3 people per 100 ha, while non-family farming employs 1.7 people per 100 ha. FRANÇA; DL GROSSI; MARQUES, 2009. The discourse fragment (DCS 9) is aligned with the official data: "To make the point, I mentioned that family farming employs 10 people per hectare, while agribusiness employs only 1 person per hectare". In addition, the teachers acknowledged that family farming is not always carried out solely by the family: *"family farming absorbs a lot of labour, not just your family, it can also have specialised labour, such as a technician or agronomist"*, also showing its capacity to generate jobs. These two excerpts end up revealing a positive representation of the characteristics of family farming, but their sense of productivity was compromised, as can be seen below.

The discourse (DCS9) contains the justification of the need for technologies used to increase productivity in agribusiness. The justification is maintained even considering *the* exclusion of people. This can be seen in the fragments: *"the needs of the amount of people there are to feed; but I commented that when you use technology you exclude labour, but productivity increases"*.

One can see in the discourse that teachers are concerned about meeting the world's demand for food. However, world food production hasn't been a problem for a long time, as currently production capacity would cater for 12 billion people, but there are seven billion people in the world and 800 million suffer from permanent hunger (ZIEGLER, 2013).

It's clear that it's not a problem of technique, that's been overcome. However, this concern can give students the idea that more intensive production is needed to feed the entire population, while in fact what we have at stake are business interests that maintain the constant demand for increased production. What really seems necessary is to reduce social inequalities and inequities by facilitating access to food. And encouraging family farming is seen as one of the most significant strategies for reducing these inequalities.

The thinking present in this discourse justifies as a "necessary evil" the exclusion of the rural man, the small property and the small producer for the sake of greater productivity. "The 21st century undoubtedly requires governance of technological innovation, but above all it requires governance of limits on the use of materials, energy and greenhouse gas emissions" (ABRAMOVAY, 2012).

In DCS 10, the passage "In my understanding, when Mrs Maria sells a lettuce, she is doing agribusiness" shows an inadequate understanding of what family farming and agribusiness are. Some teachers seem to be resistant to recognising the distinctions between these forms of production.

4 Large estates and small properties - social and economic repercussions

Four central ideas emerged from this second theme. When asked if the concentration of land by agribusiness and its consequences for the rural population had been discussed during the course, three teachers had addressed the issue and two had not.

Central Idea 1: The teacher discussed the concentration of land by agribusiness

> CSD 11 *Yes, I even showed statistical data on the number of establishments and who owns the land. The consequences, if you look at the concentration of landholdings, most of the arable land is in the hands of large landowners who grow monocultures. So, what I see is this: I don't agree when you expropriate productive land to produce just one type of product, but I don't see any problem with agricultural land that isn't being used. For example, soya in Balsas, at the beginning of the colonisation of that land, practically that land was government land that had no owner, and in order for people to seek it out, to inhabit that region, incentives were given and large areas of land were made available.*

Central Idea 2: The teacher didn't discuss the concentration of land by agribusiness

> DSC12 *No either. There wasn't because I don't work with the division of family farming and agribusiness, for me it's all one thing.*

By understanding statistical data, students can build up an understanding of the extent of land concentration. In terms of this data, according to the latest agricultural census, family farming in Brazil, despite representing 84.4 per cent of agricultural establishments, occupies an area of only 24.3 per cent of the total area devoted to agricultural activity. Although in their speeches the teachers did not express an

thought that directly considered the consequences for the rural population, this can be understood indirectly, from reflection on the data.

It would be essential to consider in class that in recent years there has been a growth in global demand for agricultural *commodities* and the consequent expansion of business interests and the land market (CARNEIRO, 2013), which has led to the occupation of this land by planting products aimed at the foreign market. Thus, critical perception could be encouraged in students, leading them to understand and reflect on the consequences of land concentration by agribusiness for the rural population.

According to Carneiro (2013), "in the case of Maranhão, this has meant encouraging the expansion of large-scale soya plantations (ANDRADE, 2007; CARNEIRO, 2013) to supply steel mills, the main effects of which will be the heating up of the land market and processes of peasant expropriation."

Therefore, it can be concluded that although the teacher works on the issue of land concentration in class with statistical data that reveals the reality of the agrarian structure, the social representation expressed by the teachers about the historical process of how this happened does not contribute significantly to the students' understanding of what, in fact, made this concentration possible and the consequences for the rural population.

In discourse DSC 12, the indistinction between family farming and agribusiness stands out, which points in the direction of not recognising different forms of occupation of space and production, stripping the concentration of land, the basis of monocultures, of its real meaning.

When asked if during the course the teacher had addressed the expansion of agribusiness and the implications of this expansion in terms of access to land for family farming, three teachers said they had and two said they had not.

Central Idea 3: The teacher spoke during the course about the expansion of agribusiness.

> CSD13 *I discussed the expansion of agribusiness. You don't see large areas of land turned over to agribusiness in all regions, it's in specific areas where that particular activity takes place, those who produce on a large scale need to increase the area. I also commented that when there isn't an adequate policy, what's left is degraded land that will need a high cost to work. And also about agrarian reform programmes.*

Central Idea 4: The teacher didn't address the expansion of agribusiness during the course.

> SCD 14 *No. And I think we need to demystify this idea that family farming is about small producers who work only for their own consumption and can't use technology.*

In CSD 13, which reads *"You don't see large areas of land turned over to agribusiness in all regions, it's in specific areas"*, this opinion reveals a misrepresentation of IBGE data which shows that family farming occupies less land in all regions of the country. The distribution of the area of family farming establishments by region is as follows: Northeast (35%), North (21%), Southeast (16%), South (16%) and Centre-West (12%) (BRASIL, 2006). The data reveals that in all regions the areas occupied by family farming establishments are smaller.

Still in the same discourse, the fragments *qqumn produces on a large scale needs an increase in area; You don't see large areas of land turned over to agribusiness in all regions, it's in specific areas"*, reveal a kind of naturalisation of the expansion of land concentration and agribusiness, following the logic of the economic model aimed at increasing exports and obtaining and maintaining a trade balance surplus, following the logic of the globalised market. I would emphasise here the importance of developing a critical stance in the pedagogical process, as well as the possibility of questioning the development model chosen by the country, which manifests itself in agriculture and the concentration of land.

> Land concentration is another important aspect of this development model, as the type of activity encouraged (livestock, soya, sugar cane, forestry plantations) is highly demanding on land resources, negatively impacting on the structure of the distribution of land ownership and possession (CARNEIRO, 2013).

The implications of this expansion in terms of access to land for family farming are revealed when they focus on the loss of land quality due to misuse and the importance of agrarian reform in this context, which strengthens the understanding of the need to rethink the ways in which land is distributed and used. Therefore, it can be seen that among the teachers who worked on the issue of the expansion of agribusiness, they dealt with the contradiction represented by the demand for land concentration by agribusiness and the need to distribute land to family farming. This contradiction could stimulate questioning approaches. It should be borne in mind that the school is a space for formulating questions, organising debates and providing diverse responses.

5. The right to food security. Food production and distribution. Inequalities, scarcity and hunger.

Five central ideas emerged from this second theme. When asked if during the course the teacher had addressed the question of "where the food that supplies the Brazilian table comes from", everyone said yes.

Central Idea 1: The teacher addressed the question "where does the food that supplies Brazilians' tables come from?".

> DSC15 *I spoke about the issue of supply for export, from planting to harvesting, and that family farming plays an important role because it contributes a lot to production. I also showed statistical data and videos, which reveal that 70% of Brazil's supply comes from family farming and the other part comes from agribusiness. I also cited the example of Maranhão, where the majority of fruit and vegetable and animal products come from abroad, which I believe is more a cultural issue than a technical one.*

The speech shows an approach that involves the stages of production *from planting to harvest* and the forms of supply for export and the domestic market, emphasising the great contribution of family farming in supplying food for the Brazilian table, including statistical data. The representation is in line with the data from the last agricultural census in 2006 in terms of the percentage of participation of family farming in the food supply.

According to the IBGE, despite cultivating a smaller area with crops, family farming is the main supplier of basic foodstuffs for the Brazilian population and despite cultivating a smaller area with pastures, family farming is an important supplier of animal protein (BRASIL, 2006).

In the same speech, there is a focus on the situation in Maranhão, where the insufficiency of food production in the state is considered to be related more to a cultural issue than to access to technology. According to Silva (2010), food production by family farming in Maranhão is one of the determining factors:

> The main idea is that, having been left in a marginal position in relation to agricultural modernisation policies, the state of Maranhão has started to pursue a model of agricultural development, the result of which is the decreasing availability of food over the years. This model is characterised, on the one hand, by the absence of an agrarian policy and, on the other, by the nature of tax incentive policies, which has resulted in increased concentration of land (SILVA, 2006).

The understanding we get from Silva's (2006) studies is that there is a lack of technical assistance; the problem does not lie with the crop, but with the socio-historical process of building the agricultural model in the state, characterised by exclusion.

When asked if during the course the teacher had considered that the lack of access to land by the rural population leads to social and health problems, three teachers said they had not worked on this issue.

Central Idea 2: The teacher considered that the

rural population's lack of access to land

leads to social and health problems.

DSC16 *I was talking mainly about social problems, such as the rural exodus, where people are expelled from their land and come to the outskirts of the cities to ask for better housing conditions, they arrive there in the invasion, and then they want roads, they want hospitals, they want schools. In addition, there's the indebtedness that comes with acquiring finance, which leads producers into depression, causing various problems.*

Central Idea 3: The teacher did not consider that the lack of access to land by the

rural population

leads to social and health

problems

CSD17 *No, because I don't think it's the lack of access to land, I think it's, specifically in Brazil, what I see, it's the lack of technical assistance, because many producers I know have the land, the lack of assistance jeopardises their profit, many are expelled from the land and go to the outskirts and health is a consequence. Not only do people not want to turn to the rural sector, unfortunately, but the companies that already have this vision and concept invest heavily.*

As for those who worked on the subject, there was a focus on social issues rather than directly on health. The discourse (DCS 16) considers policies on access to land as an instrument to guide the occupation and use of rural and urban space, which also contributes to quality of life and access to goods and services. The teacher's perception of the situation of conflict and exclusion in the countryside makes it possible to recognise inequities, which are possibly avoidable inequalities, and their relationship with the compromised health situation due to the social factors generated by exclusion.

In the discourse (DCS 17), the teacher does not deal with social and health problems as a consequence of the lack of access to land, as he considers that it is the lack of technical assistance and not access to land, as well as people's lack of interest in farming that generates these problems. It is well known that many people have land, but use archaic forms of production that end up compromising productivity, but the issue of access to land is much broader than the teacher's

50

experience, "because many producers I know have the land, but the lack of assistance compromises their profit". There is also a history marked by inequality in the distribution of land, generating conflicts over land disputes that have excluded many from the countryside.

When asked if during the course the teacher had addressed the issue of world hunger in the classroom, only one said he had not.

Central Idea 4: The teacher addressed the issue of world hunger in the classroom

> CSD18 *I discussed world hunger in class. I mentioned the issue of commodities destined for export, which are not the main food for the population, but producers are not very concerned about the issue of food in general, they just want to make a profit. I also talked about how there is enough food in the world to feed everyone, but the lack of money, especially in African countries, prevents access. It's also the case that the treaties signed at meetings of powerful countries aren't honoured, because there is an injection of money, but it doesn't reach the small producer. On top of all this, there's also the ignorance of the people, which can be seen in the waste and spoilage of fruit, for example.*

Central Idea 5: The teacher didn't address the issue of world hunger in class

> CSD 19 *I didn't talk about hunger in class*

In the discourse (DCS 18), as well as showing that the topic was worked on in class, it was worked on considering the role of capitalist agriculture in contributing to the situation of hunger in the world, as shown in the fragment "but producers are not very concerned about the issue of food in general, they just want to make a profit", as well as showing the lack of political commitment to the situation of hunger "I also talked about the stock of food that exists in the world, it would be enough to feed everyone, but the lack of money", denouncing the violation of the right to food when there is an abundance of production but inequities prevent access, thus denouncing the violation of the right to food when there is an abundance of production, but inequalities prevent access, because "of course, availability cannot be considered a guarantee of access to food, availability is a necessary condition, but not sufficient to guarantee access, which is a requirement for food security". (SILVA, 2006).

Although discourse 19 was formed by a single teacher, it is considered a Collective Subject Discourse, because it is an opinion that exists in a culture, a group, a society. Considering that the teacher is from the technical area of agriculture (food production), he possibly represents the discourse of many teachers who also don't work on the subject. Despite the fact that he hasn't worked on the subject, there's no way to consider that the topic isn't relevant, see the comic strip below:

Figure 5- Cartoon about poverty arguments

Source: Guiraldelli Jr., 2006

In this comic strip of the character Mafalda by the Argentinian cartoonist Joaquin Salvador Lavado - better known as Quino, who works with comics focusing on social and political concerns, it can be seen that Mafalda "doesn't see her mother's choking and reticence as a situation of someone who doesn't have an answer or who finds it strange to have to find an answer. On the contrary, she believes that *there is* an answer to her question" Guiraldelli Jr. (2000) when she says "I didn't *imagine that my question would be so interesting".*

It is therefore believed that agricultural schools should lead students to discuss social and political issues involving their future professional field and, above all, be prepared to provide answers that society in general is often unable to give.

6 Food safety and health

Four central ideas emerged from this second theme. When asked if the teacher had studied food safety during their training, all said yes.

Central Idea 1: Did the teacher study food safety during their training?

> CSD 20. *Yes, I studied, mainly with regard to agrochemicals, but also on this issue I saw the importance of rural producers saving their seeds so that they wouldn't be dependent on the government, as well as the importance of the quality of fruit, in addition to studying the daily quantities of food consumed, following a socio-economic survey of people in the settlement.*

The discourse (DCS 20) shows a focus on food security around the stages of production, commercialisation and consumption, based on the principles of food sovereignty and sustainability. Therefore, the discourse is aligned with the principles of the concept of SAN, which according to LOSAN.

Central Idea 2: The teacher addressed food safety issues in his lessons

CSD 21. *Yes, I did. I talked about hygienising the product so that it can be consumed reliably, and about the products that are permitted and prohibited for use in animal feed. In the fruit section, I talked about pesticide residues in fruit. As well as the importance of product certification, because a certified product is safer and is a differentiator when it comes to gaining a foothold in the market, without forgetting the importance of families' nutritional needs.*

In the discourse, the quality of the product appeared as a central point, complemented by the nutritional value of these products. It can be seen that these points are decisive for the understanding of a quality diet aimed at health, based on a diet free of agrochemicals and other products and in sufficient quantities. The discourse reveals a social representation of the way in which the issue is addressed, focusing on production and consumption techniques, without emphasising the political and social issues that compromise the population's food security.

When asked about their understanding of the relationship between access to food and the health of the population, four related it to the difficulty of accessing food and one considered the ignorance of the people as a factor that leads to a lack of good nutrition and consequently good health.

Central Idea 3: The relationship between access to food and the health of the population is related to income

CSD 22 *The relationship between access and the health of the population is due to the difficulty in accessing food due to low purchasing power, which is related to the capitalist system.*

Central Idea 4: The relationship between access to food and the health of the population is related to the ignorance of the people

CSD 23 The *relationship between access to food and health goes back to ignorance, some prefer to splurge rather than eat well, it goes back to self-awareness, level of education, it's a personal issue.*

n the discourse (DCS 22) *"low purchasing power"*, it is clear that the teachers recognise the problem of the population's low income as a factor that compromises access to food and its relationship with the exclusionary system that is capitalism. In discourse (DCS 23), the ignorance and ostentation factor appears, and may even be present in some people's lives, but the discussion goes far beyond a public that can and wants to decide whether or not to eat, it is related to unequal income distribution, the inequity present in today's world.

The representation of world hunger in the discourse (CSD23) demonstrates the failure to recognise the destructive nature in socio-economic terms of the monopolistic capitalist model of agriculture that is taking place all over the world. The perception that food security is being jeopardised by this predominant model of agriculture is therefore well undermined. In the book "Geopolitics of Hunger" by Josué de Castro, the title indicates that hunger is of political origin, and not of nature, which goes against the thought shown in the fragment: *it's a personal issue.*

At no point did they mention the capitalist model of agriculture that generates exclusion, nor that family farming is a form of production that contributes to ensuring food security and eradicating hunger. Although capitalism was mentioned, there was no comment about the current model of agriculture, whose production is organised along the lines of general production, which is capitalism, and the social problems it generates: exclusion, unemployment, hunger and, finally, compromised health.

6.3 Analysing and discussing the students' statements in the interviews

Analyses of the central ideas (CI) and discourses (DSC) enabled them to be grouped into six themes:

1. The year 2014 was declared the International Year of Family Farming by the Food and Agriculture Organisation of the United Nations (FAO);

2. Food production for agribusiness and food production by family farming.
3. Agribusiness, family farming and labour in the countryside.
4. Large estates and small properties - social and economic repercussions.
5. The right to food security. Food production and distribution. Inequities, scarcity and hunger.
6. Food safety and health.

1. The year 2014 was declared by the Food and Agriculture Organisation of the United Nations (FAO) as the international year of family farming.

Two central ideas emerged from this first theme. When the students were asked if there had been any discussion during the course about the celebration of 2014 as the International Year of Family Farming by the FAO, six of the eight students said there had been no such discussion, and two said they couldn't remember if there had been.

Central Idea 1: The student states that there was no such discussion during the course

CSD 1 *No, there was no such discussion here on the course.*

Central Idea 2: The student doesn't remember if this discussion took place during the course

The summaries of the students' statements confirmed the testimony of four of the five teachers interviewed, who said they had not worked on the subject. In general, it was clear that the students were not aware that 2014 was recognised as the international year of family farming. Considering that, even among the deliberations of the final MEC/Setec document (2009), which deals with the (Re)signification of agricultural education, which says that agricultural education institutions should prioritise the Family Farming segment and, as one of the reference elements for productive dynamics, Agroecology, (BRASIL, 2009, p. 110), it can be seen in these speeches that the students had not worked on the subject. 110), it can be seen in the students' speeches that the teachers didn't take advantage of the discussions to contribute to the training of technicians based on agroecology and sustainability, as well as failing to publicise and disseminate agroecology, points required as proposals in this document.

2. Food production for agribusiness and food production by family farming

Five central ideas emerged from this second theme. In order to obtain the answers, the students were asked how they came to see or understand food production geared towards agribusiness and food production by family farming after their training. CSD 3 was built from the speeches of seven people, while CSD 4 was built from the speeches of just one.

Central Idea 1: Agribusiness production is aimed at profit from exports, while
family farming production is aimed at local supply.

CSD 3 *I think that today I know the real importance of family farming. It's more concerned with maintaining the consumption of the farmer himself and his family, in other words, it's concerned with maintaining the nutrition of the rural man and his food. Therefore, production stays in the city itself, it's more geared towards the local sector. Agribusiness production, on the other hand, caters for the larger market, with its main focus on capital and profit. The products produced by agribusiness are generally exported. It is more export-orientated.*

Central Idea 2: Family farming generates jobs and income

CSD *A Family farming generates income for the farmer. The more we look for food, the more jobs and income it will generate for the farmer.*

The understanding of family farming as a means of survival more focused on guaranteeing subsistence was predominant in relation to the sense of generating income and work. *It is more concerned with maintaining the consumption of the farmer himself and his family,* since the understanding of the more accentuated notion of employment and income from family farming was observed in the speech of only one person in DCS 4.

Even though family farming is the main productive economic activity, even in the Northeast, revealing its potential as commercial family farming, "it still suffers from stigmas linking it to productive backwardness, relating it to subsistence" (SOUZA, 2011). It is common knowledge that small-scale family farming, due to a lack of structural and technical conditions, is limited to subsistence, but considering that official documents have revealed (IBGE/Census 2006) the potential of the family farming sector, these future technicians failed to reveal this potential in their discourse.

Only one of the eight students related the issue of employment, revealing the sector as a space for occupation and income generation. Currently, this sector is responsible for the largest number of jobs among agricultural establishments in the country, according to the latest agricultural census. From this perspective, it can be inferred that these future agricultural technicians do not see themselves as workers in this sector, since they have little connection with its employability. This inference can be confirmed when the students answered in the interview which sector absorbs more labour in the countryside, agribusiness or family farming. Five of the eight students said it was agribusiness, which is in line with the latest census data.

When the students were asked if the course had addressed the environmental impacts caused by agribusiness, they all said that it had.

Central Idea 1: During the course, the impacts on the environment caused

by agribusiness were discussed

, especially the use of pesticides.

> CSD 5: *Yes, during the course we discussed the environmental impacts caused by agribusiness. Mainly about the exaggerated use of agrochemicals in large crops, such as corn and soya. These agrochemicals contaminate the soil, groundwater and water sources, which not only harms us, but also agribusiness workers. Other impacts include the removal of native forests, riparian forests and air pollution, thus contributing to the greenhouse effect.*

The discourse reveals that the teachers were concerned to show the impacts related to the model of food production prevalent in the country, which is export monoculture based on the intensive use of industrialised agricultural inputs, with an emphasis on pesticides. This information was appropriate in the period in which the country has remained the largest consumer of pesticides since 2008 (CONSEA, 2014).

According to Abramovay (2010), livestock farming is directly responsible for the degradation of biodiversity on the planet. "In addition to the impact of farming on biodiversity degradation, it also contributes negatively to methane emissions through enteric fermentation and the management of animal waste" (BERCHIELLI, 2012).

When the students were asked if the course had covered the environmental advantages of food production by family farming, four said it had and three said it hadn't. The students were asked if the course had covered the environmental advantages of food production by family farming.

Central Idea 2: It has been commented on and the advantage is that pesticides are used on a smaller scale

> DSC6: *Yes, they talked about the advantages of family farming for the environment. Because family farmers are starting to use the agroecological way of planting, they don't use pesticides in large quantities, others don't use pesticides at all. Another advantage is that small farmers don't have to destroy so much forest in order to plant.*

Central idea: No comment.

> DSC7 : *The course didn't talk about the advantages of family farming, but I think there are advantages! One of the advantages is less use of pesticides in this form of production.*

According to Neves (2007 apud CARVALHO; MARIN 2011) "the term family farming can be used to designate a productive sector endowed with characteristics or 'values' associated with socio-environmental sustainability", but it is known that the generalisation of the term ends up hiding a diversity of these forms of production by farmers. Among the players in this sector, some use agroecological principles and others, for example, use pesticides indiscriminately, often due to a lack of technical guidance. However, the prevailing meaning when using the term family farming is to recognise it as a proposal for a sustainable form of production, and this was perceived when the students in CSD 6 related the environmental advantages of this form of production.

However, although some said they had worked on the subject, this was not evident to the other students interviewed, revealing the low impact of the content.

3. Agribusiness, family farming and labour in the countryside.

Four central ideas emerged from this second theme. When the students were asked which sector absorbs more labour in the countryside, agribusiness or family farming, and how this issue was dealt with, three students formed SCD 8, two formed SCD 9, two formed SCD 10 and one formed SCD 11. Five of the eight students believe that agribusiness is the sector that absorbs the most labour.

Central Idea 1: Agribusiness was dealt with during the course.

CSD 8 *I think it's agribusiness. This subject was barely touched on during the course. The teacher touched on it. It's agribusiness that absorbs the most labour in the countryside, because it's large-scale production, so it needs more jobs than family farming, because it needs people to work with machines, such as machine operators, as well as others, such as technicians, engineers, etc.*

Central Idea 2: Agribusiness, but it wasn't discussed during the course.

CSD 9 I think it's agribusiness, but this subject wasn't covered during the course

Central Idea 3: Family farming and what was covered in the course

CSD 10 *I think it's family farming. This issue was covered in the course, but not much. Family farming absorbs more labour because it requires manual work, an example of which is that the machine does the work of 10 or 15 men.*

Central Idea 4: Family farming, but it wasn't dealt with in the course

CSD 11 *I think it's family farming, right? But this issue wasn't dealt with in the course.*

Considering that teachers' work has a direct influence on learning, and that when asked if this issue was dealt with in the classroom, the students expressed themselves in different/divergent ways, it can be seen that there was no uniformity in the teachers' discourse in the classroom.

This could be seen in the speeches of the students who said they had been taught about this issue in class. One speech says: "*Agribusiness absorbs the most labour in the countryside, because it is a large-scale production, so it needs more jobs than family farming, because it needs people to work with machines, such as machine operators, as well as others, such as technicians, engineers.*" This speech goes against the grain of the lack of opportunities generated by agribusiness in the countryside.

Teachers' actions should lead to the construction of knowledge, or even to the development of understanding and socialisation.

recognised by the students are still confused and sometimes unfounded, manifesting a naivety or lack of a political and critical vision.

Others, through their discourse, understood THE issue adequately: "*Family farming absorbs more labour because it requires manual work, an example of this is that a machine can do the work of 10 or 15 men*". However, their opinions were still fragile: "*I think it's family farming, right?*"

4. Large estates and small properties - social and economic repercussions

Six central ideas emerged from this second theme. When asked what reasons or needs "justify" the large concentration of land by agribusiness, seven students formed SCD 12 and only one formed SCD 13.

Central Idea 1: Large-scale production for profit

> CSD 12 *I think agribusiness needs a large expanse of land because of the huge investment made in large-scale production. As production increases, so does the concentration of land, so they can make more profits, which is all they want.*

Central Idea 2: Lack of land distribution.

> CSD 13 *I think the concentration of land is due to a lack of land distribution to small farmers.*

All the students had a clear perception of the power of agribusiness to concentrate land. However, only one person commented on the lack of land distribution to family farmers as a "justification" for the reasons for land concentration, thus characterising a more political-critical view of the lack of agrarian reform.

The same didn't happen in discourse DCS 12 (statement by the majority), because the passage *"I think that agribusiness needs a large area of land because of the huge investment"* shows that the students received the question without being surprised, i.e. the question was naturalised among them. From this we can see that these students' perception of the environment in relation to the appropriation and use of natural resources - particularly land - is anthropocentric.

Among the categories established for the different representations of the environment, according to Reigota (1991) "globalising: - highlights the reciprocal relationships between nature and society; anthropocentric: - favours the usefulness of natural resources for human survival, and naturalistic: - emphasises only the natural aspects of the environment".

Since the school is a social environment capable of getting to know the students' environmental perception and raising their awareness of the environment, it has a great chance of achieving success in the process of citizenship education aimed at socio-environmental responsibility and building a critical political vision, which has not been widely perceived when it comes to issues of land concentration.

When asked if there had been discussions about the concentration of land by agribusiness and its consequences for the rural population, four students said there had not been such a discussion, while four said there had been, considering issues of rural exodus and contamination by pesticides.

Central Idea 3: There was no such discussion.

Central Idea 4: Yes. Rural exodus and pesticide contamination.

Still on the subject of land concentration, now focussing on its consequences for the rural population, in the statements made by the four students who said that the subject had been covered in the course, there was no mention of the government's responsibility in terms of land reform so that small producers could have access to land.

What is clear, as the following extract reveals, *The purchase of land from small producers by large agribusiness companies happens, and they are almost forced to sell their land, causing a rural exodus,* is that there is a logic in which there are those who can afford to buy, and those who seem to have no other choice.

According to Silva, Agrícola and Pietrafesa (2010), within the agrarian question, there is a conflict that is the result of the confrontation between the territory of the peasantry and the latifundia and agribusiness, and the idea of the statement in DCS 15 does not reveal this conflict, in other words, the theme does not seem to have been problematised in such a way as to provide a deeper understanding of the social issues that encompass the topic.

When asked if there had been any discussion during the course about the expansion of agribusiness and its implications for family farming in terms of access to land, five students said they had not dealt with the subject during the course and three said they had.

Central Idea 5: The expansion of agribusiness was not addressed during the course

Central Idea 6: The expansion of agribusiness and its consequences for
family farming were discussed

Silva, Agrícola and Pietrafesa (2010) analysed the relationship between Brazilian agribusiness, its colonial heritage and the impositions placed on it by the economic model and revealed that "the search for productivity and the expansion of industrial and agricultural capital, at all costs and above all else, continue to be the goals of this type of economic activity, social and environmental problems tend to grow and increase their consequences." (SILVA; AGRÍCOLA; PIETRAFESA, 2010).

Therefore, this expansion must be dealt with in the classroom, and this discussion must go beyond showing a situation without reflecting on its causes and consequences. In the excerpt *Yes, it was discussed that with this expansion, small producers have to sell their land to meet the needs of large producers*. Once again, the problematised, questioning nature of the situation does not appear, just a naturalisation of what is imposed.

5. The right to food security. Food production and distribution. Inequities, scarcity and hunger.

Nine central ideas emerged from this fifth theme. When asked if the question of "where the food that supplies Brazilians' tables comes from" had been addressed during the course, seven said it had and only one said it had not.

Central Idea 1: Yes, during the course it was discussed where the food that supplies
Brazilians' tables comes from

.

> CSD 18 *Yes, this issue was addressed. Food comes from the big producers of the big agribusiness companies, and I've always thought that. The teachers said that most of what we eat comes from family farming. The small farmer consumes the food he produces and contributes to the local supply, fuelling our table. I remember there was this approach to annual crops.*

Central Idea 2: No, this issue was addressed

> CSD 19 *It wasn't.*

Studies carried out by the IBGE show that family farming is responsible for most of Brazil's food production. Based on this data and its relationship with the discourse generated by the students (those who said they had worked on this topic in class), despite the fact that the teachers had portrayed this reality, considering family farming as a major food producer, *through the teachers' comments, they said that the majority of what we consume comes from family farming,* there is still a misconception in the perception of some students that food comes from agribusiness, the idea that is present is that this sector is powerful in its production capacity, but that in fact what is produced is not always going to be transformed into food, perhaps this is the misconception that justifies the phrase *Food comes from the big producers of the big agribusiness companies, and I've always thought that.*

When the students were asked whether the mode of access to land (landowner, sharecropper, employee, etc.) had been considered as an important factor in guaranteeing family farmers' income and how it had been dealt with, three students said it had not been dealt with and five said it had.

Central Idea 3: This was not discussed.

CSD 20: *This issue wasn't addressed much*

Central Idea 4: In a way, this issue was dealt with.

CSD 21: *In a way, it was touched on. I remember the government's incentives supporting rural people to stay working there and not leave their land and go hungry in the big cities. I also did some research on Pronaf, the federal government's programmes. I think all this is important for family farming to stay in the countryside, because if they don't stay in the countryside there won't be any money. They talked about it, but I don't remember exactly. They also talked about how this business is done, how it's sold and how it's negotiated. Before, I didn't even know what a sharecropper was, I didn't even know there was such a thing as renting land.*

According to NEY and HOFFMANN (2009) "in Brazil, in particular, the main structural determinants of income disparity in the primary sector found in the literature are the *distribution of wealth,* especially land ownership." Therefore, what will determine income to a greater or lesser degree will be the form of land ownership. Considering that the squatter exploits land that is not his, the sharecropper makes an agreement with the landowner to divide production and the landowner is the owner, it is clear that income will vary according to the farmer's situation.

When students research PRONAF, they learn about the policies aimed at small producers who are in the most diverse situations of ownership. To be a beneficiary of the programme, one of the requirements is that they exploit a plot of land as an owner, squatter, tenant, lender, partner, PNRA concessionaire or public land permit holder (FAQ, 2014). It is possible that through activities such as this research programme, students will understand how land ownership is important for guaranteeing family farmers' income.

When asked if during their training they had considered that the lack of access to land by the rural population leads to social problems and if these problems have consequences for the population's health, one student formed DSC 22, one student formed DSC 23, and six students formed DSC 24.

Central Idea 5: No, during the course the lack of access to land by the rural population led to social problems. And I believe that this situation does not interfere with the population's health.

CSD 22 *During the course, they didn't say that the rural population's lack of access to land causes social problems, but I think it does, because without land to work on, there will be a lack of food. But I don't know if this affects the population's health, I don't think it does at all.*

Central Idea 6: No, during the course the lack of access to land by the rural population led to social problems. But I believe that this situation generates problems for the population's health.

CSD 23 *No, during the course the lack of access to land by the rural population led to social problems. Personally, I think it does, social problems, yes, and it also causes health problems, for example, hunger.*

Central Idea 7: Yes, during the course it was considered that the lack of access to land by the rural population leads to social problems. And I know that this situation generates health problems for the population.

CSD 24: *Yes, during the course it was considered that the lack of access to land for the rural population leads to social problems. Because when small producers can't produce on their own land, they tend to go to the big city, because without land how are they going to support themselves? Then, in the cities, they go to live in unsuitable places, without sanitation, because even though they go in search of a job, without qualifications, they can't find one, and all this poverty affects their health. And what happens? If you don't eat well, you end up in hospital.*

Although only one student didn't relate the problem to men's health due to lack of access to land, this reveals that he doesn't relate poverty to *health, as seen in his speech: But I don't know if this affects the health of the population, I don't think it does anything.* It's worrying that a speech by a student over the age of 18 who is completing technical training reveals such a lack of correlation between social and health factors. However, the vast majority relate the effects of conflict in the countryside and rural exodus to hunger, which is the predominant factor in compromising health.

According to Sauer and Leite (2012), "growing investments in land assets threaten food security and sovereignty because they further concentrate agricultural production in a few commodities, favouring monopolies in food and agro-energy production." This highlights the danger of the expansion of agribusiness in compromising health through food insecurity, and the importance of addressing this issue in the classroom in order to encourage students to have a critical perception.

When asked what justifies the hunger of almost 1 billion people in the world when food production is sufficient to supply the entire world population, seven students formed SCD 25 and four formed SCD 26.

Central Idea 8: Hunger occurs because of inequality in the world

> CSD 25: *I think hunger is caused by the capitalist system, which creates inequalities. The consequence of this is the formation of rich and miserable countries, which suffer from poor food distribution because they don't have the resources to obtain it. The prejudice generated by the rich towards the poor means that food exports favour the richest. I think that hunger occurs due to a lack of financial conditions, which compromises people's income in order to buy food. I think that hunger occurs due to a lack of investment by the governments of poor countries themselves. One example is what happens to small farmers who lose ground to agribusiness, leading them to stop farming and look for jobs in the big cities. This shows a lack of investment in food security.*

Central Idea 9: Hunger occurs because of food waste

> CSD 26: *I think that hunger is mainly due to the great waste of food, even though Brazil has plenty of food, we don't know how to use it.*

The students' perceptions of hunger were moderately positive. There was a tendency among the students to attribute hunger to poor income distribution, i.e. inequalities related to the capitalist system, and the government's responsibility to solve the problems of poverty and social exclusion. In a study carried out in Portugal (ESTUDO SOBRE A PERCEPÇÃO DA POBREZA, 2011), which sought to find out the population's perceptions of poverty, one of the findings was that a large proportion of those surveyed said that the government was responsible, in other words, they "extemalised" the responsibility and, ultimately, like these students, may not consider themselves to be part of the solution.

6. Food safety and health.

When asked whether they had been taught about food safety and the focus given to this issue, all the students formed SCD 27.

Central Idea 1: The topic of food safety was addressed by SESC

> DSC 27: *There was a SESC week here at the school and they talked about food safety. SESC was here at the school, but not everyone was able to take part because there was a limit on enrolment. I learnt about the importance of peeling the food we throw away.*

Food security is one of the basic rights of the human person and includes the guarantee of access to safe, quality food, in sufficient quantity and on a permanent basis, without compromising access to other essential needs and based on healthy eating practices.

In the discourse above, the students' perception of the concept of Food Security is limited to ways of utilising food or alternative forms of feeding. However, based on other discourses in this study, generated by the students and the contents involving Agroindustrial Production, Animal Production, Plant Production, and Planning and Management (PLANO DE CURSO, 2009) present in the course plan and which allow direct contact with the production of food of animal and vegetable origin, these students do have basic knowledge of food safety, what they have difficulty with is relating the concept of FS to practices.

As in the study by RANCATTI; LASTA;, M. L. (2014), which aimed to assess the knowledge and perceptions of food safety among agricultural technical course students in Paraná, it can also be said that despite the curriculum not having a specific subject on food safety, the students have basic knowledge on the subject RANCATTI; LASTA;, M. L (2014), as this was evident in the other speeches.

When the students were asked about the relationship between access to food and the health of the population, seven students formed SCD 28 and one SCD 29.

Central Idea 1: The relationship between access to food and the health of the population is the

consumption of

quality food.

CSD 28 *I think that if you have access to a good diet, at the right times, and this food is healthy, of good quality, you will have good health and a good quality of life. But if the food is contaminated, for example, carrots with mites or full of pesticides, this can lead to many illnesses.*

Central idea 2: The link between access to food and health is income.

CSD 29 *I think that when you can't afford to buy what's recommended, it affects your health.*

In their speeches, most of the students' perception of the population's health was more focused on food quality. Following this reasoning, for these students, possibly one way of solving the problems generated by the relationship between access to food and health would be to provide healthy food free of chemical and biological contaminants.

This understanding is correct, but not complete, as it partly covers the principles of food security, which are quality and quantity (related to health), but does not reveal the social aspects of access, which also compromise health. However, when the key question related to "access", only one student mentioned the need for income in order to have access to food and guarantee health I think that when you can't afford to buy what is recommended, it affects your health.

4.4 Analysis and discussion of the teachers' statements in the focus group

When the teachers were asked to comment on the use of transgenic seeds, the contamination of soil and water resources by family farming and agribusiness, they formed DSC 01.

Central Idea 1: Transgenics don't pollute.

> DSC 1 *Transgenic does not pollute, transgenic in the way they are producing the seed has no problem at all. The imbalance comes from monoculture! Where eucalyptus trees are used, nothing grows, it's a repellent, it's a natural repellent, it doesn't give insects, it doesn't give anything, there are no birds, there's no pollination, there's nothing, there's just eucalyptus, which has a tremendous environmental impact. For example, sugar cane monoculture has had an impact on the Atlantic rainforest. Yes, yes! Let me talk about transgenics: the big problem is that it doesn't obey federal legislation. Then there's the spacing for the natives, then there's a problem with pollination, so the big issue is that it doesn't obey the legislation and that influences it, do you understand? So that's the big controversy in agriculture with transgenics. There was a caterpillar called Helicoverpa, wasn't there? There isn't a poison in the world that can fight it.*

According to Ribeiro and Marin (2012) "the controversial issue of GMOs involves divergent opinions where, on the one hand, sectors of society strongly defend the extensive use of this technology, while, on the other hand, sectors repudiate its use without first carrying out a long and detailed study of the impacts that planting and consuming these foods could have on human health and the environment." Economic and political interests justify these opposing opinions on the use of transgenics. "This controversy involves various actors, such as scientists, farmers, environmentalists and government representatives, referring to the level of uncertainty attributed to these foods, in relation to so-called 'food security'". (CAMARA; MARINHO; GUILAM; NORDARI, 2009)

In the teachers' statements, there is a clearly defined position in defence of the use of transgenics, and at no point does it reveal any uncertainty or discussion about the safe viability of the application of transgenics for food security, as shown in *the* passage *"Transgenics the way they are producing the seimenle has no problem whatsoever"*. The statement reveals that the problem is not in the use of transgenics, but in the way it is being used, in monocultures that don't follow the legislation regarding spacing, which ends up reducing biodiversity or expanding the monoculture. The genetically modified product itself seems to be safe to them, even at a time of so many discussions in academic circles questioning this "safety", academies of which these professors are a part, as they are part of a federal public teaching, research and extension institution.

4.5 Analysis and discussion of the students' statements in the focus group

When the students were asked to comment on Foucault's phrase: "Every system of education is a political way of maintaining or modifying the appropriation of discourses, with the knowledge and powers they bring with them. "

Central Idea 1: Education changes the way we see the world

> CSD 01 *The course here has opened up a thousand and one frontiers and opportunities for me. I think that the way of thinking, the way of expressing oneself for those leaving the course, for those finishing now, will be better as technicians. A lot of things, for me in particular, I never thought would be like this, like working with an animal, I'd already done it, but in a technical way, like, in theory I didn't know how to work in agriculture. And the area of the course is a very promising area and has a very large job market, and not just jobs, but also becoming an employer, a small producer. Producers don't realise that they have to study to be able to produce better and better at low cost and high quality. I've already learnt that.*
>
> *I learnt a lot here, a lot that I never even imagined was like this. I used to say that I didn't eat farmed chicken because it contained too many hormones. Today I know the truth! I now know that in order for a chicken to have a short slaughter period of up to a month, a certain amount of genetic work has gone into it. It's a supplemented ration, it's a ration with a lot of care...*
>
> *I used to think that they put hormones in chickens to make them grow faster. And that would harm us. But today, so much so that I can't eat a chicken that's raised in the countryside, because a thousand and one things come to mind. Today, I've studied balanced nutrition and I've got all the* assistance of a vet there, so I'm not going to eat rubbish. In the past I used to eat pork, now I don't even care about pork, I only eat pork because I know where it comes from and what kind of diseases that animal can have. I can eat it, I can catch it. Now, a frozen chicken from a farm already has a certain level of sanitary care. And it's also its diet, it's not going to eat rubbish to reach that weight...
>
> And I think it's interesting because often when you don't have the knowledge, sometimes you think one thing but it's not what you think, isn't it? Then when you study, you have knowledge and you learn that it's totally different from what you imagined, so it's very interesting to study in order to learn and really get your mind around what's true and what's real.
>
> In my opinion, the farming course is one of the best because I was brought up in the area, my whole family was brought up in the area. Education does change concepts, it changes your conscience, it changes the way you listen, speak and see your environment. So that was very productive, my course, my farming course.

In 2014, in order to assess knowledge and perceptions of food safety, a study was carried out with students on the agricultural technical course at the Assis Brasil State Professional Education Centre, located in the municipality of Clevelândia-Paraná, with the aim of training students in food safety and hygiene practices in the food industry. The results revealed the students' prior knowledge of food safety and hygiene practices in the food industry. The target audience was students on the technical course integrated into secondary education, totalling approximately 280 students. (RANCATTI, LASTRA, 2014)

The previous perception of food safety in the above-mentioned study revealed that factors linked to food poisoning, shelf life, hygiene and health were important for guaranteeing food safety in the industry, as ways of reducing the risk of food-borne illnesses. Although our study is not specifically focused on food safety in industry, but rather on the social issues of food safety, in this CSD 01, in which the students emphasise the importance of the course, it is clear that, when it comes to the issue of food safety, the major focus is also on issues linked to food poisoning, shelf life, hygiene and health, as ways of reducing the risk of illness, as in the Paraná study. This focus on food safety is correct, but limited because it does not take into account political, economic and social aspects, so that the student would have a critical perception of the production sector in which they will be inserted.

This is justified by the IFMA's course plan, which proposes as a professional profile, in accordance with the competences and skills developed, a number of points, such as: safely implementing, managing and monitoring new forms of animal husbandry according to productivity, developing an entrepreneurial spirit, carrying out and managing agro-industrial activities, acting as a promoter and disseminator of products, knowing the health standards and procedures for marketing products and their derivatives (COURSE PLAN, 2009).

Another study aimed to analyse the social representations of students and graduates of the technical course in agriculture at the Federal Southeast Institute of Minas Gerais, Barbacena Campus, about the agricultural labour market, as well as to identify their main problems and potentials during their academic life and after graduation, when they were already in the world of work. The results showed that they are faced with a clash of ideas and values that they bring from their life experiences. It was observed that few young people were focused on the world of rural labour, even though the region was predominantly family farming, and this revealed a need for pedagogical restructuring and adapting the course to the regional rural reality (OLIVEIRA,2011).

Making a comparison with the Barbacena Campus study, IFMA students may also be being trained with a focus very much on agro-industry, when the state of Maranhão is eminently agricultural, and the vast majority of agricultural establishments are family farms, This could lead to the same conflicts in the future as the students in the aforementioned study, especially as there are students who already have families who live off small-scale production and aim for entrepreneurship "In *my opinion, the agriculture and livestock course is one of the best because I was brought up in the area, my whole family was brought up in this area. "*

It's hard to produce as a small producer with your family when you think, "But today, so much so that today, I can't eat a chicken that's raised in the countryside, because a thousand and one things come to mind. Today I've studied and I know what types of diseases that animal can have. I can eat it, I can take it. Now, a frozen farm chicken already has a certain sanitary care". Dias (2014) deals with the growth of chickens from other aspects. In her article "Para ir além do alimento-mercadoria" she considers that the food industry in recent years has reduced food to consumption (commodity) and has not overcome hunger, threatening health, as can be seen below:

> The disproportionate growth of the chickens compromises the humidity of the meat, so injections of brine and other substances guarantee a natural appearance. The industry's entire endeavour is to look homemade, artisanal and natural, as if it were freshly made at home. In the name of this naturalness, the health of plants, animals, soil and humans can be threatened (DIAS, 2014).

It is clear from the students' statements that teaching should be more geared to the economic and social context in which they live, given that the demand for courses in the agricultural area is preferably from those who already have a history of rural production in the family, so that students could value rural production practices more and identify with them.

4.6 Analysing didactic content

Based on the theoretical foundation and the results of the interviews, the focus of this analysis was to learn about the policy for training agricultural technicians, with an emphasis on the curriculum, subjects and general content of the course for agricultural technicians at IFMA. In this way, the aim was to relate this knowledge to the perception of the final-year students on the themes of economic development, agribusiness expansion, family farming, food production and access, food security and health.

The technical course in Subsequent Farming had 1560 hours and lasted two years. With regard to the justification for this course, it aims to keep up with technological transformations, enabling a solid scientific education, where there is a harmonious and creative interaction with nature, always seeking through research, new discoveries or alternatives to improve its intervention in the productive field. (COURSE PLAN, 2009). Also as a justification:

> And because Maranhão is an eminently agricultural state, where the geographical situation, the history of occupation, the demographic indicators and the educational indicators mean that the population of Maranhão currently tends towards new technologies and a new professional qualification profile for this new moment, thus aiming to create professional education courses that meet market demand. (COURSE PLAN, 2009)

The aim of the syllabus is for students to acquire competences, logical reasoning, skills and behavioural dispositions that will enable them to enter the modern world in a more productive, critical and creative way. The objectives of the plans include:

> - To train a qualified professional within the precepts of skills and competences to act with dignity in the world of work;
> - To provide technical scientific knowledge within a critical and reflective attitude of citizenship in the exploitation of agricultural areas;
> - Keeping abreast of production market trends in the agricultural field, valuing sustainable development as a priority goal for achieving progress;
> - Enable the development of teaching practices that meet the needs of the labour market and guarantee the development of workers' personal aptitudes and aspirations. (COURSE PLAN, 2009)

The plan proposes to prioritise service to the state's production chain and reaffirms its commitment to the teaching-learning process in terms of professional training aimed not only at doing, but also at the student entering the labour market knowing how to do and why to do it, emphasising and valuing the human being as a citizen who is aware of their place in society. (COURSE PLAN, 2009)As for the student's training itinerary, this involves Planning and

Administration of Plant Production, Planning and Administration of Animal Production, Planning, Management and Agroindustrial Production and the Curricular Internship. This takes place as follows:

Table 03: Curriculum matrix for the medium-level technical professional education course in agriculture and livestock farming

QUALIFICATION	UNIT	DISCIPLINES	Total
Plant Production	Agriculture I	Olericulture Gardening and Landscaping Agroecology	90 40 40
	Agriculture II	Annual Crops Agricultural Mechanisation Topography	90 60 60
	Agriculture III	Fruit growing Irrigation and Drainage Construction and Rural Facilities	90 60 60
SUB-TOTAL			570h
Animal Production	Zootechnics I	Poultry Farming / Laying Small Animals	90 80
	Zootechnics II	Pig farming Sheep and goat farming	90 60
	Zootechnics III	Dairy and beef cattle farming Forage farming	90 40
SUB-TOTAL			450h
Planning and Management	Management	Agricultural Management I Agricultural Management II Fundamentals of Information Technology Planning and Projects	40 40 40 40
SUB-TOTAL			160h
Agro-industrial Production	Management	Agroindustrial Production Technology	80
SUB-TOTAL			80h
Agricultural Techniques		Egg Techniques Meat Techniques Fish Techniques Dairy Techniques	120
Curricular Internship			180
TOTAL WORKLOAD			1.560h

Table 04: Qualification: Medium-level Agricultural Technician

Functions	Sub-functions
1 - Planning and Projects	1.1 Vocational studies of the region
	1.2 Project Development
2 - Plant Production	2.1 Land use and management capacity
	2.2 Study of climatic factors and their relationship with plants
	2.3 Plant growth and development
	2.4 Propagation and planting
	2.5 Pest, disease and weed management
	2.6 Drawing up the harvest and post-harvest plan
3 - Animal Production	3.1 Animal Reproduction
	3.2 Genetic Improvement
	3.3 Animal nutrition and forage
	3.4 Breeding management
	3.5 Animal health

	3.6 Production sourcing and preparation
4 - Planning and Management	4.1 Setting up and monitoring the project's administrative structure
	4.2 Drawing up the property's operating plan
	4.3 Monitoring the commercialisation process
	4.4 Monitoring, control and evaluation of the production process.

As for the course curriculum, it is organised in accordance with the National Curriculum Guidelines for Professional Education - DCNEP/99. The guidelines are orientations for thought and action, and end up becoming an instrument for the political direction of student training. Among the guiding principles of these DCNEP is the development of competences for employability. This principle is taken into account in the course plan when it is defined:

> This course is linked to competences and skills, with flexibility in keeping up with technological innovations and the constant search for updates in a globalised society. Competence is strictly linked to the world of work; when you develop the application of skills, knowledge, abilities and attitudes, you increase the possibility of carrying out a professional activity on the economic scene (PLANO, 2009).

Considering that this research is dealing with an agricultural school, it should be borne in mind that an agricultural school is not the same thing as vocational education in the countryside, so in general terms the vocational education courses at agricultural schools are not designed to train farmers, but they do face a scenario in which two aspects predominate: one is the preparation of salaried workers for agro-export companies and the other is the training of extension workers - linked to public bodies or even companies - to provide technical assistance to farmers. This scenario is the challenge to be faced by this education.

This challenge is even greater when we look at the current scenario in Maranhão, a place where there is a strong tendency for agro-industry and agribusiness to grow, but where 91 per cent of agricultural establishments are family farms, which requires Agricultural Technicians with extension skills. It is believed that one of the great challenges when it comes to the labour market in Maranhão is to show that this training of extension workers is what will enable students to enter the world of work.

After reading and analysing the course plan, we realised that the focus given to the training of technicians was geared towards serving the set of businesses related to agriculture and livestock from an economic point of view, i.e. geared towards agribusiness and agro-industry practices. This was due to the plan's concern with rapid technological changes and the need to meet the demands of the Maranhão market with the growth of these activities.

The National Curriculum Guidelines for Vocational Education - DCNEP /99 that guided the course had as their guiding principles for technical vocational education those set out in article 3 of the LDB, plus the following: I - independence and articulation with secondary education; II - respect for aesthetic, political and ethical values; III - development of skills for workability; IV - flexibility, interdisciplinarity and contextualisation; V - identity of the professional profiles at the end of the course; VI - permanent updating of courses and curricula; VII - autonomy of the school in its pedagogical project. It also had the following criteria for the organisation and planning of courses: I - meeting the demands of citizens, the market and society; II - reconciling the demands identified with the vocation and institutional capacity of the school or education network. (Brasil. CNE/CEB, 1999).

Considering the principle of "developing competences for workability" and that competences are strictly linked to the world of work, in 2010 a debate arose on updating the guidelines for technical vocational education at secondary level, the initial procedures of which were two public hearings organised by the National Education Council (CNE), which resulted in the document National Curriculum Guidelines for Technical Vocational Education at Secondary Level in Debate - Text for Discussion.

In the debate there was criticism of the competency-based curriculum, because it was understood that schools have always developed competencies, but when the question of the competency-based curriculum is raised, the only training possible is training, which implies the selection of knowledge orientated predominantly towards functional performance. In these discussions, it was understood that in this way the reference to the school's characteristic properties, which are culture and basic systematised scientific knowledge (technical and technological), is lost.

In this way, the course plan complied with the guidelines it was meant to follow and because of the plan's concern with the rapid technological changes and the need to meet the demands of the Maranhão market, we can understand why the results of the analyses of the students' speeches generated by the individual interviews and the focus group were so different. It can be considered that the agricultural teaching of the course in question was strongly centred on learning the techniques needed to meet the needs of the promising agribusiness and agro-industry market, and the extensionist character was not prominent. And even though one of the course's objectives was to "Provide technical scientific knowledge within a critical-reflective attitude of citizenship in the exploration of agricultural areas", this left something to be desired, as can be seen from the excerpts from the DCS below, generated by the students in the survey:

> DSC8 "It's *agribusiness that absorbs the most labour in the countryside, because it's large-scale production, so it needs more jobs than family farming, because it needs people to work*

> *with machines, such as machine operators, as well as others, such as technicians, engineers, etc. "*

> CSD7 *"*The course *didn't talk about the advantages of family farming,* but I think there are advantages! One of the advantages is less use of *pesticides in this form of production."*

> CSD16 *"No, it wasn't discussed. I don't remember us talking about it... "*

> CSD 18 *"Food comes from the big producers, the big agribusiness companies, and I've always thought that."*

> CSD 22 *"During the course they didn't say that the lack of access to land for the* rural *population* causes social problems, but I think it does, because without land to work on there will be a lack of food. But I don't know if this affects the population's health, I *don't* think *it does at all."*

> CSD 23 *"No, during the course the lack of access to land by the rural population led to social problems. Personally, I think it does, social problems, yes, and it also leads to health problems, for example, hunger."*

> CSD 26 *"I think that hunger is mainly due to the great waste of food, even though Brazil has a lot of food, we don't know how to use it."*

> CSD / Focus Group *"I've learnt a lot here, a lot that I never even imagined was like this, like .... I had a speech that I didn't eat.*
> *farm chicken because it had too much hormone. Today I know the truth! I now know that in order for a chicken to have a short slaughter period of up to a month, a certain amount of genetic work has gone into it. It's a supplemented ration, it's a ration with a lot of care.   In the past, I used .... to think that they put hormones in the hen to keep her healthy.*
> *chicken grow faster. And that would harm us. But today, so much so that I can't eat a chicken that's raised in the countryside, because a thousand and one things come to mind. "*

Although the DCNEP/99 had its limitations, it still allowed for "permanent updating of courses and curricula" and "autonomy of the school in its pedagogical project", thus failing to prioritise the principles proposed by the 2009 MEC document dealing with the (Re)signification of agricultural education, a document that originated from the need to rethink the prevailing model, taking into account the transformations in society and production processes. However, it is known that more than what is proposed in the plan or the guidelines, teaching practice (in the classroom) is fundamental, because it is the teachers, the professionals who are "at the cutting edge" who end up having the greatest responsibility in the process of training students, as it is possible for them to be "at the cutting edge".

have the best plans and guidelines, if you don't have professionals committed to critical-reflective training, nothing will happen.

Thus, the knowledge and perceptions of the final-year students on the themes of economic development, the expansion of agribusiness, family farming, food production and access, food security and health from a critical-reflective point of view were not well achieved, leaving something to be desired, as they often saw uncritical acceptance of imposed values. According to the results, the influence of market logic on the training policy for students in technical vocational education at secondary level can be seen. Considering that there is currently a new DCNEP/2012, it is more involved in terms of its social and environmental responsibilities than the previous one, but still adjusted to partnerships with the private sector, this can be seen in the criticism made when comparing the two guidelines:

> In the case of the new National Curriculum Guidelines for Vocational Education and the opinions and resolutions that gave them their normative character (2011/2012), when compared with their previous version (1999) and also compared with the proposal for the DCNEP made by the Social Collective, they reveal that maintaining fragmented and disjointed paths hurts the principles of integration claimed for the polytechnic school. In addition, the strong call for training geared towards the interests of the labour market signals the persistence of the school's tangency with the interests of the business community and the deepening of the social division of labour (BERNARDIM; SILVA, 2014).

It should be noted that vocational education in Brazil needs to take on the idea of work as an educational principle, as it is opposed to a reductionist, utilitarian, atrophying and essentially restrictive view of human formation (CIAVATTA, 1992).

CHAPTER 7

CONCLUSION

The aim of this study was to identify the perceptions of teachers and students on the technical agricultural course regarding the promotion of knowledge and reflection on the relationship between the way land is occupied and used (expansion of agribusiness, intensive use of pesticides, incentives for family farming) and the population's food security.

In view of the situations addressed involving the themes of economic development, agribusiness expansion, family farming, food production and access, food security and health, this research showed that many of the teachers and future agricultural technicians, according to their personal perceptions, demonstrated prior knowledge of the effects of agribusiness expansion (within the model of capitalist agriculture) on socio-environmental aspects and the impacts on family farming. It was noted that although these issues had been addressed by some of the teachers in the classroom, the students provided information in an isolated and fragmented way, revealing a lack of commitment to training critical and reflective professionals in their field.

Although most of the people interviewed were aware of the importance of the role of family farming in keeping people in the countryside and its advantages in relation to the environment, it was noted that, in general, this form of production is seen as backward. During the course, the focus on agro-industrial production outweighed the agro-ecological approach. The result of the students' perception was a view that the greatest productivity in terms of food and job creation permeated the agribusiness sector, the result of an education that placed little value on food production by family farmers.

With regard to the study of food safety, which is at the centre of this research, it was found that teachers are generally concerned about the stages of food production, focusing on hygiene, sanitation, equipment and trained staff to guarantee food safety. This concern is also perceived by the students, who find it difficult to accept more artisanal forms of production as safe for consumption, which is very common among family farmers. From this, we can see that during the educational process there were elements that favoured the development of a vision that was more focused on technique, leading them to see and interpret the world in terms of the logic of the labour market.

As a result, we realise that the professional training of agricultural technicians has left something to be desired when it comes to topics that address the issue of food security with a focus on the social impacts caused by the predominant model of agricultural production in Brazil. This prevents future professionals from taking a critical view of this production model, which generates social inequalities and could jeopardise their role as rural development agents for family farmers. Based on the results, it is suggested that in order to overcome this fragmented training, it would be necessary to think specifically about this state and the family farming that needs these professionals, a training programme that manages to articulate three axes: family farming, food security and agroecology.

To this end, I propose that in order to improve the didactic content present in the course plan that governed the training of these students, it could undergo changes and one of them would be to place Agroecology as a cross-cutting theme in all disciplines and not as an isolated discipline, since it is a science that requires a systemic approach, another issue to be suggested is that there should be a discipline that prepares students to work with public policies, such as PAA, Pronaf, as well as working with the Rural Environmental Registry, so that the student would have a more global vision. It is essential that the school provides the capacity for students and teachers to always be willing to appropriate new knowledge with a view to training for citizenship, seeking to make individuals capable of thinking and acting responsibly in order to promote practices aimed at improving the quality of life.

Therefore, the importance of an agricultural school in the state of Maranhão working in teaching, research and extension is due to the possibilities of training rural development agents who can contribute to the social transformation of a state marked by poverty, especially in rural areas. Among these marks is a state where the incidence of poverty reaches 53.38% and the prevalence of food insecure households is over 50%. Considering that Maranhão is an eminently agricultural state and the Brazilian state with the highest percentage of the population living in rural areas, with 87% of the rural labour force employed in family farming establishments, it is essential that the state's agricultural schools are committed to transforming the reality in which they find themselves, seeking to face the challenges imposed by agricultural education through its (Re)signification.

REFERENCES

ABRAMOVAY, Ricardo. Food versus population: is the Malthusian ghost resurfacing? **Cienc. Cult.** [online], vol.62, n. 4, 2010.

. Inequalities and limits should be at the centre of Rio+20. **Estud. av.** vol.26 no.74 SãoPaulo   , 2012. Disponívelem : http://www.scielo.br/scielo.php?script=sci_arttext&pid=SO 103-40142012000100003. Accessed on: 12 June 2015.

ALVÂNTARA, Anelise Montanes; VESCE, Gabriela Eyng Possolli. **Social representations in the discourse of the collective subject in the context of qualitative research,** 2008. Available at: http://www.pucpr.br/eventos/educere/educere2008/anais/pdf/724_599.pdf. Accessed on: 12 June

2015.

ANDRADE, Maranhão and Piauí have less than half the population with guaranteed food, **Pnad.**
Rio de Janeiro, 2013. Available at: <http7/noticias.uol.com.bi/cotidiano/ultimas-
noticias/2014/12/18/maranhao-e-piaui-tem-menos-da-metalfda-populacao-com-alimentacao-
garantida.htm>. Accessed on: 12 June 2015.

AZAR, Zaira Sabry. **Land concentration as the centrality of the agrarian question in
Maranhão.** In: 5th International Conference on Public Policies. São Luís: UFMA, 2011.

BARBOSA, Leila Cristina Aoyama; ZANON, Angela Maria. Environmental approach in technical
education curricula: an analysis of the agricultural technician courses at the Federal Institute of
Mato Grosso (IFMT). In: National seminar on professional and technological education.
Proceedings... Belo Horizonte: CEFET/MG, 2010.

BARBOSA, Leila Cristina Aoyama. Environmental approach in technical education curricula: an
analysis of the agricultural technician courses at the Federal Institute of Mato Grosso. **Anais...**
Campo Grande: Federal University of Mato Grosso do Sul, 2010. Available at:
http://www.senept.cefetmg.br/galerias/Anais_2010/Artigos/GTl/ABORDAGEM_AMBIENT
AL.pdf>. Accessed 12 May, 2014.

. **Environmental perception of future agricultural technicians:** the vision of students on a
technical course about the environment in which they live. Rio Grande Federal University of Mato
Grosso do Sul. Campo Grande: UFMS, 2010. Available
at:http://sistemas.ufmt.br/ufint.evento/files/8728ceed-5e9c-4b07-afl e-9530639d0042.doc. Accessed
12 May, 2014.

BELIK, Walter. Brazil's food and nutrition security policy: conceptions and results. In: **Food and
nutrition security.** Campinas, n. 2, v. 19, 2012. Available
at:<http://www.unicamp.br/nepa/arquivo_san/volume_19_2_2012/19-2_artigo-8.pdf>. Accessed
on: 18 November 2013.

. Perspectives for food and nutritional security in Brazil. **Saúde soe.,** v.12, n.l, p.12-20, jun., 2003.
Available at:
<http ://www. scielo .br/scielo.php?script=sci_arttext&pid=SO1041290200300010000
4&lng=en&nrm=iso>. Accessed on: 22 October 2013.

BERCHIELLI, Telma Teresinha; MESSANA, Juliana Duarte; CANESIN, Roberta Carrilho.
Enteric methane production in tropical pastures. **Rev. Bras. Saúde Prod. Anim.** [online]. vol.13,
n.4, 2012. Available at: <http://dx.doi.org/10.1590/S1519- 99402012000400010>. Accessed on:
15 May 2015.

BERNARDIM, Márcio Luiz; SILVA, Mônica Ribeiro. **Curricular Policies for Secondary
Education and Vocational Education: proposals, controversies and disputes in the face of
the proposals in the Conae 2014 Reference Document.** Available at: <
http://www.jpe.ufpr.br/nl6_3.pdf>. Accessed on: 07 Jul.2015.

BRAZIL, Ministry of Education. National Programme for Access to Technical Education and
Employment. Technological axis: natural resources / agricultural technician. **National Catalogue
of Technical Courses,** 2012. Available at:
<http://pronatec.mec.gov.br/cnct/et_recursos_naturais/t_agropecuaria.php>. Accessed on: 18
November 2013.

. Secretariat of Professional and Technological Education - SETEC. **(Re) significificação do Ensino Agrícola da Rede Federal de Educação Profissional e Tecnológica.** (Final Document). Brasília DF, 2009.

. National Commission on Social Determinants of Health (CNDSS). Open letter to the presidential candidates. **Revista Rades Comunicação e Saúde.** September 2006. Available at: <http:www.determinantes.fiocruz.br Accessed on: 18 November 2013.

. National Food Security Council. **Human right to adequate food,** 2013. Available at: <http://www2.planalto.gov.br/consea/o-conselho/conceitos-   l/human-right-to-adequate-food>. Accessed on: 18 November 2013.

. National Food Security Council. **Food Security and Food Sovereignty,** 2013. Available at: <http://www2.planalto.gov.br/consea/o-conselho/conceitos- lAccessed on: 18 November, 2013.

. National Food Security Council. **National Food and Nutrition Security System,** 2003. Available at: <http://www2.planalto.gov.br/consea/o- conselho/conceitos-l/sistema-nacional-de-seguranca-alimentar-e-nutricional>. Accessed on: 18 November 2013.

. Law No. 11.346, of 15 September **2016.Organic Law on Food and Nutritional Security,** 2014. Available   at:<http://www.planalto.gov.br/ccivil_03/_ato2004-2006/2006/lei/ll   1346.htm>. Accessed on: 18 November 2014.

. Ministry of Education National Education Council. National Curricular Guidelines for Technical Professional Education at Middle Level. PARECER CNE/CEB N°: 11/2012 COLLEGIATE: CEB APPROVED ON: 9/5/2012.

. **Saf-Creditorural,** 2015. Available at: http://www.mda.gov.br/sitemda/secretaria/saf-creditorural/sobre-o-programa>. Accessed on: 18 November 2013.

. Ministry of Agrarian Development. **Family farming in Brazil and the agricultural census,** 2006. Available at: <file :///C :/U sers/ Cliente/Downloads/ Censo%20da%20 Agricultura%20familiar.pdf.> Accessed on: 18 November 2014.

. CEB RESOLUTION NO. 4, OF 8 DECEMBER 1999, 1999. Available at:<http://portal.mec.gov.br/cne/arquivos/pdf/rceb04_99.pdf>. Accessed on: 18 November 2014.

BUARQUE, Sérgio C. **Metodologia de planejamento do desenvolvimento local e municipal sustentável.** Brasília: IICA, 1998.

BURLANDY, Luciene. Conditional cash transfer and food and nutrition security = Conditional cash transfer programme sandfoodand nutrition security. **Ciência & Saúde Coletiva.** Rio de Janeiro: UFF, v. 12, n.6,2007. Available at: <http://www.scielo.br/scielo.php?pid=S 141381232007000600007&script=sci_arttext>... Accessed on: 18 November 2013.

BUSS, Paulo Marchiori; PELLEGRINI FILHO, Alberto. Health and social determinants. **PHYSIS Revista Saúde Coletiva.** Rio de Janeiro, v. 17, n. 1, 2007. Available at: <http://www.scielo.br/pdEphysis/vl7nl/vl7nla06.pdf>. Accessed on: 18 November 2013.

CAMARA, Maria Clara Coelho; MARINHO, Carmem L.C.; GUILAM, Maria Cristina Rodrigues and NODARI, Rubens Onofre. Transgenics: evaluation of possible food (in)safety through scientific production. Hist. cienc. saude-Manguinhos [online]. 2009, vol.16, n.3, pp. 669-681. Available at < http://www.scielo.br/scielo.php?pid=S0104-59702009000300006&script=sci_arttext> Accessed on: 01 Jun. 2015.

CAPRA, Fritjof. **The web of life.** São Paulo: Saraiva, 1996.

CARNEIRO, M. D. S. . **Land, labour and power:** conflicts and social struggles in contemporary Maranhão. 1. ed. São Paulo: Annablume, v. 1,2013.

POLITICAL LETTER, III National Meeting of Agroecology. Bahia, 2014. Available at: file:///C:/Users/user/Downloads/Carta%20Politica_ENA%20(l).pdf . >. Accessed on: 18 November 2014.

CARVALHO, Simone Pereira de; MARIN, Joel Orlando Bevilaqua. Family farming and the sugarcane agroindustry: social impasses. **Rev. Econ. Sociol. Rural** [online]. Vol.49, n.3, 2011. Available at: < http://dx.doi.Org/10.1590/S0103-20032011000300007.>. Accessed on: 13 Apr., 2015

CIAVATTA FRANCO, Maria. **Labour as an educational principle for children and adolescents.**

Tecnologia Educacional, ABT, Rio de Janeiro, 21 (105/106):25-29, Mar./Jun. 1992

SOCIAL COLLECTIVE. **National Curriculum Guidelines for Vocational Education**

Technical High School in debate. Text for discussion. Brasília. 2010. Available at:

https://www.google.com/url?sa=t&rct=j&q=&esrc=s&source=web&cd=l&ved=OCCOQFjAA

&url=http%3A%2F%2Fportal.mec.gov.br%2Findex.php%3FItemid%3D%26gid%3D6695%

26option%3Dcom_docman%26task%3Ddoc_download&ei=6ji3UdXXHKSU0QH3 60GAB

w&usg=AF Qj CNESICwOTT z_Qg J2RlvQgoLvX9MWw&sig2=T gsq7 ao2veKIMV

mWMnM3g&bvm=bv.47 534661,d.dmQ. Accessed on: 08 July 2015.

CONSEA. **Principles and Guidelines for a Food and Nutrition Security Policy.** Reference Document for the Second National Conference on Food and Nutrition Security. Brasilia: CONSEA, 2014.

DEPONTI, Cidônea Machado; ALMEIDA, Jalcione. **On the process of social mediation in development projects:** a theoretical reflection. Porto Alegre, 2008.

DIAS, Juliana. Going beyond food as commodity. **Other Words.** São Paulo, 2014. Available at: <http://outraspalavras.net/posts/para-ir-alem-do-alimento-mercadoria>. Accessed on: 13 April 2015.

STUDY ON THE PERCEPTION OF POVERTY in Portugal, 2011. Available at: http://www.amnistia-intemacional.pt/files/Relatoriosvarios/RelatorioPobreza_com_indice.pdf >. Accessed on: 13 April 2015.

FAO. Food and Agriculture Organisation. Agricultural data base, 2003. Available at: <http://www.fao.org.>. Accessed on: 12 May 2014.

. **What is family farming?** International Year of Family Farming, 2014. Available at: http://www.fao.org/family-farming-2014/home/what-is-family-farming/pt/>. Accessed on: 12 May, 2014.

FAQ. **National Programme for Strengthening Family Farming** - Pronaf. Are
Paulo: Central Bank of Brazil, 2014. Available at:
http://www.bcb.gov.br/pre/bc_atende/port/PRONAF.asp#2 .> Accessed on: 13 Apr., 2015.

FARIA, Mauro Velho de Castro. **Evaluation of environments and products contaminated by pesticides.** In: PERES, Frederico; MOREIRA, Josino Costa. E veneno ou é remédio, agrotóxicos, saúde e ambiente. Rio de Janeiro: FIOCRUZ, 2003.

FAVACHO, Fernando; SOUSA, Ronier. **Professional training for agricultural technicians with an emphasis on agroecology:** current situation and challenges at the Federal Agrotechnical School of Castanhal - Pará. Pará, 2006. Available at: <http://www.cultura.ufpa.br/cagro/pdfs/TextoN017_A_formacao_profissional_de_tecnico_e m_agropecuaria.pdf>. Accessed on: 03 Jan. 2014.

FRANÇA, Caio Galvão de; DL GROSSI, Mauro Eduardo; MARQUES, Vicent P.M de Azevedo. **The 2006 agricultural census and family farming in Brazil.** Brasília: MDA, 2009.

FREITAS, Carlos Machado de; GARCIA, Eduardo Garcia. Labour, health and the environment in agriculture. **Rev. bras. Saúde ocup.,** São Paulo, v. 37, n.125, 2012.

GHIRALDELLIJR, P. **O que é filosofia da educação.** 2ª ed. Rio de Janeiro: DP&A, 2000.

GIANEZINI, Miguelangelo et al. Geotechnology applied to agribusiness: concepts, research and supply. **Revista Economia & Tecnologia (RET)** v. 8, n. 2,2012.

GOLDENBERG, Mirian. **The art of research.** Rio de Janeiro: Record, 2004.

GOMES. Heloisa Maria; MARTINS, Hiloko Ogihana Marins. **Teaching action in vocational education.** São Paulo: Senac São Paulo, 2004.

GUANZIROLI, Carlos, et. al. **Family farming and agrarian reform in the 21st century.** Rio de Janeiro: Garamond, 2009.

IBGE. **National Household Sample Survey:** Food Security, 2013.
Available at:
<http://www.ibge.gov.br/estadosat/temas.php?sigla=ma&tema=pnad_seguranca_alimentar_2 013.>. Accessed on: 18 November 2014.

. São Luís: summary of information2010 Available at:
<fonte:http://www.cidades.ibge.gov.br/xtras/temas.php?lang=&codmun=211130&idtema=1 6 &search=||s%EDntese-das-informa%E7%F5es>. Accessed on: 12 September, 2014.

IFMA. **General guidelines for technical and higher education at the Federal Institute of Maranhão,** São Luís, 2011. Available at:
<http://www.ifina.edu.br/proen/arquivos/Legislacao/diretrizes_gerais_do_ensino_ifina.pdf>. Accessed on: 12 September, 2014.

FEDERAL INSTITUTE OF EDUCATION, SCIENCE AND TECHNOLOGY OF MARANHÃO (IFMA). Course Plan. São Luís, 2009.

LEFEVRE, Fernando; LEFEVRE, Ana M. **Depoimentos e discursos:** uma proposta de análise em pesquisa social. Brasília: Líber Livro, 2005.

. **Qualitative research taken seriously,** São Paulo, 2003. Available at:
http://www.fsp.usp.br/~flefevre/Discurso_o_que_e.htm>. Accessed on: 18 November 2013.

. The collective subject that speaks. **Interface** (Botucatu) vol.10 no.20 Botucatu
Jhttp://www.scielo.br/scielo.php?script=sci_arttext&pid=Sl 414-32832006000200017uly/Dec.,
2006. Accessed on: 18 November 2013.

MALUF R. S. et al. **Food safety booklet.** Available at
<http://www.forumsocialmundial.org.br/download/tconferencias_Maluf_Menezes_2000_por. pdf>.
Accessed on: 18 November 2013.

Convivialist MANIFESTO: declaration of interdependence. São Paulo: Annablume, 2013.

MARANHÃO. INSTITUTO DEFERAL DE EDUCAÇÃO, CIÊNCIA E TECNOLOCIA DO
MARANHÃO. **General Guidelines for Technical and Higher Education at IFMA.** São Luís,
2011. Available at:

http://www.ifma.edu.br/proen/arquivos/Legislacao/diretrizes_gerais_do_ensino_ifina.pdf>.
Accessed on: 03 Jan. 2014.

MINAYO, M. C. S. & SANCHES, O. Quantitative-Qualitative: Opposition or
Complementarity? **Cad. Saúde Públ.,** Rio de Janeiro, year 3, 1993.

MINAYO, Maria Cecília de Souza (org.) **Social Research.** Theory, method and creativity. 18
ed. Petrópolis: Vozes, 2001.

MOREIRA, Magda Regina Santiago. A look at family farming, human health and the
environment. **Cienc. Cult. [online].** vol.65, n.3, 2013.

NEY, Marlon Gomes; HOFFMANN, Rodolfo. Education, land concentration and income
inequality in rural Brazil. **Rev. Econ. Sociol. Rural [online],** vol.47, n.l, 2009. Available at: <
http://dx.doi.org/10.1590/S0103-20032009000100006>. Accessed on: 03 Jan. 2015.

OLIVEIRA, José Alcir Barros de. **The Social Representations of Students and Graduates of
the Technical Course in Agriculture at the Federal Institute of Southeast Minas Gerais
Barbacena Campus on the Agricultural Labour Market.** Dissertation (Master's Degree in
Agricultural Education). Rio de Janeiro: Agronomy Institute, Federal Rural University of Rio de
Janeiro, Seropédica, 2011.

PERES, F., MOREIRA, J. C. & Luz, C. The impacts of pesticides on health and the
environment. **Ciênc. Saúde Coletiva,** v.12, 2007.

PIGATTO, Giane Magrini; LINK, Diomsio. Environmental education and food safety in the
context of medium-level technical vocational education in agriculture. **Electronic Journal on
Environmental Management, Education and Technology,** v.4, n.4, 2011.

PINHEIRO, Anelise Rizzolo de Oliveira. Reflexões sobre o processo histórico/Político de
Construção da Lei Orgânica de Segurança Alimentar e Nutricional. **Food and nutritional
security.** Campinas, v. 2, v. 15, 2008. Available at:
<http://www.bvsde.paho.org/texcom/nutricion/historico.pdfi>. Accessed on: 18 November
2013.

RANCATTI, Aline; LASTA, Daiane. **Food safety and hygiene practices in the food industry:** training farm-school students. 2014. 37 f. Course Conclusion Work (Graduation) - Federal Technological University of Paraná, Pato Branco, 2014. Available at:<http://repositorio.roca.utfpr.edu.br/jspui/handle/l/3376>. Accessed on: 18 November 2014.

REIGOTA, M.A.S. The environment and its representations in education in São Paulo, Brazil. **Uniambiente.** Brasília, year 2, 1, 1991.

RIBEIRO, Isabelle Geoffioy; MARIN, Victor Augustus. The lack of information on Genetically Modified Organisms in Brazil. **Ciênc. saúde coletiva [online],** vol.17, n.2, 2012.

RIBEIRO JÚNIOR, José Amalado dos Santos; CASTRO JÚNIOR, Raimundo Campos; BOTELHO, Raimundo Edson Pinto. **Uneven development of capitalist modernisation in the territory of Maranhão:** discursive theory, productive (re)arrangements and socio-environmental conflicts. Ourinhos, v. 6, n. 2, 2012.

RODRIGUES, Sávio José Dias; ALENCAR, Francisco Amaro Gomes de. Considerations on the appropriation of space in Maranhão by soya agribusiness: the ideology of development and land accumulation in the Gerais de Balsas micro-region. **Boletim Goiano de Geografia,** [S.I.], v. 31, n. 1, august, 2011. Available at: <http://www.revistas.ufg.br/index.php/bgg/article/view/15396/9580>. Accessed on: 03 Jan. 2014.

SACHES, Ignacy. **Back to the visible hand:** the challenges of the Second Earth Summit in Rio de Janeiro. Estud. av. vol.26, São Paulo, 2012.

SANTOS, B. de S. (org). **Producing for a living:** the paths of non-capitalist production. Rio de Janeiro: Brazilian Civilisation, 2002.

SILVA, Brunna Angélica Evarista da ; SILVA, Maria Madalena da. **Building New Paradigms for Agricultural Education:** Introductory Notes. Vi Congresso Norte Noreste de pesquisa e inovação. Tocantins, 2012. Available at:< file:///C:/Users/Client/Downloads/1362-13781-l-PB.pdf>. Accessed on: 03 Jan. 2014.

SILVA, José de Ribamar **Sá. Food security, family agricultural production and agrarian reform settlements in Maranhão.** São Luís, 2006. Available at:< http://www.tedebc.ufma.br/tde_busca/arquivo.php?codArquivo=38. Accessed on: 03 Jan. 2014.

SILVA, Alexandre Rodrigo Choupina Andrade; AGRÍCOLA, Josie Melissa Acelo; PIETRAFESA, José Paulo . **Brazilian land structure: conflicts, exclusion and environmental damage in national biomes.** In: VIII Latin American Congress of Rural Sociology, 2010. Available at:<http://www.alasru.org/wp-content/uploads/201 1/07/GT7-Alexandre- Rodrigo-Choupina.pdf.>. Accessed on: 03 Jan. 2014.

SAUER, Sérgio; LEITE, Sérgio Pereira. Agricultural expansion, prices and land grabbing by foreigners in Brazil. **Rev. Econ. Sociol. Rural [online].** vol.50, n.3, 2012. Disponível em:<http://dx.doi.org/10.1590/S0103-20032012000300007> . Accessed on: 03 Jan. 2015.

SCHMIDHUBER, Josef. Global food security under climate change. **PNAS,** v.104, n. 50, p. 19703-19708, dec. 2011. Available at: <http://www.pnas.org/content/104/50/19703.full>. Accessed on: 19 October 2013.

SCHNEIDER, Sérgio. Situating rural development in Brazil: the context and issues under debate. **Revista Económica Política,** v. 30, n. 3, jul-set,/ 2010. Available at:<

http://www.uffgs.br/pgdr/arquivos/773.pdf>. Accessed on: 18 November 2013.

SOUZA, Luciano Ricardio de Santana. The conservative modernisation of Brazilian agriculture, family farming, agroecology and pluriactivity: different ways of understanding and constructing the Brazilian countryside. **Cuad. Desarro. Rural [online].** Vol.8, n.67, 2011.

ZIEGLER, Jean. **Mass destruction: the geopolitics of hunger.** São Paulo: Cortez, 2013.

APPENDICES
APPENDIX 1 - TOPICS COVERED IN THE STUDENTS' INTERVIEWS

The Food and Agriculture Organisation of the United Nations (FAO) declared 2014 the **International Year of Family Farming.**

Food production for agribusiness and food production by family farming.

Agribusiness, family farming and labour in the countryside.

Large estates and small properties - social and economic repercussions.

The right to food security. Food production and distribution. Inequalities, scarcity and hunger.

Food safety and health.

The following questions were intended to stimulate reflection and feedback:

1. The year 2014 was declared by the Food and Agriculture Organisation (FAO) as the international year of family farming. Was this discussed at any point during the course? In what way?
2. Considering your knowledge during your training in the agricultural sector, what reasons would have led the UN to hold such a celebration?
3. From your training, how did you come to see or understand food production geared towards agribusiness and food production by family farming?
4. During the course, was the question "where does the food that supplies the table of Brazilians come from" addressed? How was this question addressed?
5. In your opinion, which sector absorbs more labour in the countryside, agribusiness or family farming? Was this issue addressed during the course? In what way?
6. It is well known that agribusiness has a high concentration of land. What reasons or needs "justify" the high concentration of agribusiness land?

7. Were there any discussions during the course about the concentration of land by agribusiness and its consequences for the rural population? Please comment on these consequences.

8. Because Brazil has historically been marked by the concentration of land and income, has this favoured the large farmers of Brazilian agribusiness to receive privileges? What are these privileges?

9. Was the expansion of agribusiness discussed during the course? During this discussion, did you comment on the implications of this expansion, in terms of access to land, for family farming? What implications were considered?

10. During your training, did you consider the method of access to land (owner, sharecropper, employee, etc.) as an important factor in guaranteeing family farmers' income? In what way?

11. During your training, did you consider that the lack of access to land by the rural population leads to social problems? What are these problems? Do these problems have consequences for the population's health? What are they?

12. Were food safety issues addressed during your training? What was the focus on this issue?

13. During the course, were the impacts on the environment caused by agribusiness addressed? Can you comment on these impacts?

14. During the course, were the advantages for the environment of food production by family farming discussed? Please comment on these advantages.

15. Considering the fact that the world today has enough produce to supply the entire world population, how would you justify the hunger of almost a billion people in the world?

16. What is the relationship between access to food and the health of the population?

17. Is there any discussion of food safety at any point during your internship? Where is your internship taking place?

APPENDIX 2 - TOPICS COVERED IN TEACHER INTERVIEWS

The Food and Agriculture Organisation of the United Nations (FAO) declared 2014 the **International Year of Family Farming.**

Food production for agribusiness and food production by family farming.

Agribusiness, family farming and labour in the countryside.

Large estates and small properties - social and economic repercussions.

The right to food security. Food production and distribution. Inequalities, scarcity and hunger.

Food safety and health.

The following questions were used to stimulate reflection and feedback:

1. The year 2014 was declared by the Food and Agriculture Organisation of the United Nations (FAO) as the international year of family farming. Was this discussed at any point during the course? In what way?
2. Why would the UN have organised such a celebration?
3. How do you describe food production for agribusiness and food production for family farming?
4. You work on this subject in your course. What kind of reflection do you get students to have on this subject?
5. During the course, was the question "where does the food that supplies the table of Brazilians come from" addressed? How was this question addressed?
6. During the course it was discussed which sector absorbs more labour from the countryside, agribusiness or family farming? In what way?
7. Were there any discussions during the course about the concentration of land by agribusiness and its consequences for the rural population? Please comment on these consequences.

8. Because Brazil has historically been marked by the concentration of land and income, has this favoured the large farmers of Brazilian agribusiness to receive privileges? What are these privileges?
9. Was the expansion of agribusiness discussed during the course? During this discussion, did you comment on the implications of this expansion, in terms of access to land, for family farming? What implications were considered?
10. During your classes, did you consider access to land as an important factor in guaranteeing family farmers' income? In what way?
11. Does access to land alone guarantee family producers' income? In addition to access to land, what else would be needed for these producers to strengthen their integration into society and the market, and consequently their income?
12. Does the rural population's lack of access to land lead to social and health problems? What are these problems?
13. Have you covered food safety issues in your classes? What focus did you give to this topic?
14. Did your classes address the environmental impacts caused by agribusiness? Can you

comment on these impacts?

15. In your lessons, did you discuss the environmental advantages of family farming for food production? Please comment on these advantages.

16. Considering the fact that the world today has enough production to supply the entire world population, due to its enormous production capacity, how would you justify the hunger of almost a billion people in the world? Have you ever raised this issue in your classes?

17. What political and economic factors maintain hunger in the world? Have you addressed these issues in your classes? In what way?

18. What do you think is the relationship between access to food and the health of the population?

19. What is your professional background? Did you study food safety during your professional training? At what point? What focus did you give to this issue? Did you discuss it in class? In what way?

APPENDIX 3 - TOPIC ADDRESSED IN THE STUDENTS' FOCUS GROUP

1. "Every education system is a political way of maintaining or modifying the appropriation of discourses, with the knowledge and powers they bring with them."[1] . What are your thoughts on this sentence?

APPENDIX 4 - THEME ADDRESSED IN THE TEACHERS' FOCUS GROUP

1. Comment on the use of transgenic seeds.

APPENDIX 5 - MANDATORY SUBMISSION FORM - STUDENT'S FREE AND INFORMED CONSENT FORM (TCLE)

I would like to invite you to take part in the research on **"Training of agricultural technicians and their perception of the relationship between land occupation and use and the population's food security".** This research has Ana Cláudia Caminha de Melo as a student/researcher and is under the guidance of professor/researcher Elvira Maria Godinho de Seixas Maciel. IFMA's technical course in agriculture and livestock in the subsequent modality caters for students who have already completed secondary school and are studying at IFMA for their technical vocational training. You have been selected because you are a student on the subsequent technical course in agriculture at IFMA-Campus Maracanã.

---

[1]FOUCAULT, Michel. The Order of Discourse. Translated by Laura Fraga de Almeida Sampaio. São Paulo: Loyola, 1996

The aim of this study is to identify, observe and learn how knowledge and reflection on the relationship between land occupation and use and the population's food security is promoted. This research will be carried out by interviewing students and teachers at IFMA - Campus Maracanã and analysing the didactic content offered in the technical training course in agriculture at IFMA. The research is a dissertation developed in the Master's programme in Public Health and the Environment at the National School of Public Health, Fiocruz.

Your participation in this research will consist of talking to the researcher about your academic activity and what you think about the relationship between the use and occupation of land and the population's food security, what led you to think this way and the contribution of the training offered by the IFMA to your current understanding of the subject. This conversation will take place after the research project has been presented at the school to the students and teachers, and the names and contact details of people who would like to take part in the research as volunteers will be requested, and the research participants will be recruited so that a date can be set for the focus group and interview. The conversation will take place in two stages: through your participation in an interview at which only you and the researcher will be present, and in a focus group. Both stages will be recorded, with your prior authorisation, but the information obtained through this research will be confidential and we will seek to ensure your anonymity (both in the recording and the transcription).

Researcher's initials: ___________________________

Participant's heading:___________________________

The data will be disclosed in such a way as not to identify you. However, situations may be described in the study that may allow you to be identified by a person who knows you, but, as stated above, we will do our best to minimise this risk. The recordings of the conversations and the written and transcribed material will be kept by me for a period of five years before being destroyed, <u>and will only be used in this research</u>. The maximum duration of the interview and focus group is estimated at two hours. At the end of the research, all material obtained during the research will be kept on file for at least 5 years, as stipulated by Resolution 466/2012 of the National Health Council and the guidelines of the Research Ethics Committee of the National School of Public Health (CEP/ENSP).

The results of the research will be publicised in lectures aimed first at the participants and then at the general public. As the research is part of a master's programme, articles will be published and a dissertation will be presented to a panel of professors who are experts in the field. The benefit of your participation will be to contribute to better training in vocational education. The results of the research could be used by academic and/or political circles to define and implement agricultural, environmental, health and, above all, educational policies. The possible risk related to your participation is embarrassment at not feeling comfortable answering the questions. In this case, it is clear that your participation is not compulsory. You can withdraw your consent at any time without any harm being caused by your refusal.

This ICF will be written on three pages and in two copies and you will receive one of them. No

The term sheet contains the telephone number and institutional address of the principal investigator and the REC, so you can ask any questions you may have about the project and your participation now or at any time. If you have any questions about the ethical conduct of the study, please contact the Research Ethics Committee of the National School of Public Health (ENSP/FIOCRUZ), located in Rio de Janeiro. The Ethics Committee is the body that aims to defend the interests of research participants in their integrity and dignity, contributing to the development of research within ethical standards.

Researcher's initials:_______________________

Participant's heading: _______________________

The ENSP/FIOCRUZ Ethics Committee can be contacted at:

Telephone and fax - (OXX) 21- 2598-2863; e-mail: cep@ensp.fiocruz.br; website:

http://www.ensp.fiocruz.br/etica

Address: Sérgio Arouca National School of Public Health/ FIOCRUZ, Rua Leopoldo Bulhões, 1480 -Térreo - Manguinhos - Rio de Janeiro - RJ - CEP: 21041-210

Ana Cláudia Caminha de Melo - Researcher
Telephone: (98) 8907-0123; email: anaclaudia@ifma.edu.br

I declare that I have understood the objectives and conditions of my participation in the research. I authorise the recording of the interview and focus group and agree to take part in the research.

Participant's signature

Researcher's initials:

APPENDIX 6 - MANDATORY SUBMISSION FORM
TEACHER'S INFORMED CONSENT FORM (TCLE)

I would like to invite you to take part in the research on **"Training of agricultural technicians and their perception of the relationship between land occupation and use and the population's food security".** This research has Ana Cláudia Caminha de Melo as a student/researcher and is under the guidance of professor/researcher Elvira Maria Godinho de Seixas Maciel. IFMA's technical course in agriculture and livestock in the subsequent modality caters for students who have already completed secondary school and are studying at IFMA for their technical vocational training. You have been selected because you are a teacher on the subsequent technical course in agriculture at IFMA-Campus Maracanã.

The aim of this study is to identify, observe and learn how knowledge and reflection on the relationship between land occupation and use and the population's food security is promoted. This research will be carried out by interviewing students and teachers at IFMA - Campus Maracanã and analysing the didactic content offered in the technical training course in agriculture at IFMA. The research is a dissertation developed in the Master's programme in Public Health and the Environment at the National School of Public Health, Fiocruz.

Your participation in this research will consist of talking to the researcher about your academic activity and what you think about the relationship between the use and occupation of land and the population's food security, what led you to think this way and the contribution of the training offered by the IFMA to your current understanding of the subject. This conversation will take place after the research project has been presented at the school to the students and teachers, and the names and contact details of people who would like to take part in the research as volunteers will be requested, and the research participants will be recruited so that a date can be set for the focus group and interview. The conversation will take place in two stages: through your participation in an interview at which only you and the researcher will be present, and in a focus group. Both stages will be recorded, with your prior authorisation, but the information obtained through this research will be confidential and we will seek to ensure your anonymity (both in the recording and the transcription).

Researcher's initials:_________________________

Participant's heading: ______________________

The data will be disclosed in such a way as not to identify you. However, situations may be described in the study that may allow you to be identified by a person who knows you, but, as stated above, we will do our best to minimise this risk. The recordings of the conversations and the written and transcribed material will be kept by me for a period of five years before being destroyed, <u>and will only be used in this research</u>. The maximum duration of the interview and focus group is estimated at two hours. At the end of the research, all material obtained during the research will be kept on file for at least 5 years, as stipulated by Resolution 466/2012 of the National Health Council and the guidelines of the Research Ethics Committee of the National School of Public Health (CEP/ENSP).

The results of the research will be publicised in lectures aimed first at the participants and then at the general public. As the research is part of a master's programme, articles will be published and a dissertation will be presented to a panel of professors who are experts in the field. The benefit of your participation will be to contribute to better training in vocational education. The results of the research could be used by academic and/or political circles to define and implement agricultural, environmental, health and, above all, educational policies. The possible risk related to your participation is embarrassment at not feeling comfortable answering the questions. In this case, it is clear that your participation is not compulsory. You can withdraw your consent at any time without any harm being caused by your refusal.

This ICF will be written on three pages and in two copies, and you will receive one of them. The form contains the telephone number and institutional address of the principal investigator and the REC, so you can ask any questions you may have about the project and your participation now or at any time. If you have any questions about the ethical conduct of the study, please contact the Research Ethics Committee of the National School of Public Health (ENSP/FIOCRUZ), located in Rio de Janeiro. The Ethics Committee is the body that aims to defend the interests of research participants in their integrity and dignity, contributing to the development of research within ethical standards.

Researcher's initials: ________________________________

Participant's heading:________________________________

The ENSP/FIOCRUZ Ethics Committee can be contacted at:

Telephone and fax - (OXX) 21- 2598-2863; e-mail: <u>cep@ensp.fiocruz.br; website:</u>

<u>http://www.ensp.fiocruz.br/etica</u>

Address: Sérgio Arouca National School of Public Health/ FIOCRUZ, Rua Leopoldo Bulhões, 1480 -Térreo - Manguinhos - Rio de Janeiro - RJ - CEP: 21041-210

Ana Cláudia Caminha de Melo - Researcher
Telephone: (98) 8907-0123; email: anaclaudia@ifma.edu.br

91

I declare that I have understood the objectives and conditions of my participation in the research. I authorise the recording of the interview and focus group and agree to take part in the research.

Participant's signature

Researcher's initials:_______________________
Participant's heading: _______________________

ANNEX 1 - Letter of Consent

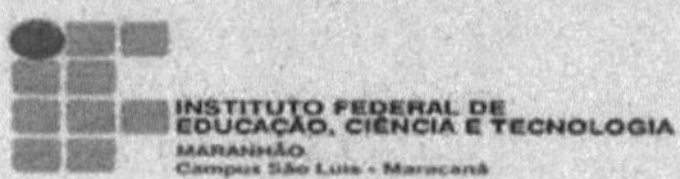

CARTA DE ANUÊNCIA

A Direção Geral do Campus São Luís – Maracanã – Instituto Federal de Educação, Ciência e Tecnologia do Maranhão – IFMA, declara anuência em relação ao projeto intitulado **"Formação do Técnico em Agropecuária e sua Percepção acerca da relação entre o modo de ocupação e o uso da terra e a segurança alimentar da população. "** (programa de mestrado interinstitucional em Saúde Pública e Meio Ambiente da Fundação Oswaldo Cruz –FIOCRUZ/IFMA), desenvolvido pela servidora Ana Cláudia Caminha De Melo, Professora, com lotação no IFMA Campus São João dos Patos , e autoriza a realização de atividades de pesquisa e aplicação de entrevistas e grupos focais junto à comunidade estudantil e docentes do IFMA Campus Maracanã.

Outrossim, informamos que a pesquisadora devera cumprir legislações pertinentes à ética das pesquisas e das atividades acadêmicas desenvolvidas no IFMA Campus Maracanã.

São Luís, 26 de junho de 2014

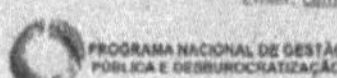

Lucimeire Honorius Castro
Diretora Geral
Instituto Federal - MA
Campus São Luís - Maracanã
Portaria nº 4.371 - DOU de 13.09.2012

Av. dos Curiós S/N – Vila Esperança – São Luís- MA, CEP. 65.095-460 Cx. Postal 433. Fones: (98) 3311 8585/8504 - Fax- (98) 3311 8504
Email: campusmaracana@ifma.edu.br Site: http://www.ifma.edu.br

PROGRAMA NACIONAL DE GESTÃO
PÚBLICA E DESBUROCRATIZAÇÃO
Organização adesa desde setembro de 2004

Printed by Books on Demand GmbH, Norderstedt / Germany